Landscapes and Lives

Landscapes and Lives

Environmental Dispatches on Rural India

Mukul Sharma

OXFORD

UNIVERSITY PRESS

OXFORD
UNIVERSITY PRESS

YMCA Library Building, Jai Singh Road, New Delhi 110001

Oxford University Press is a department of the University of Oxford. It furthers
the University's objective of excellence in research, scholarship, and education
by publishing worldwide in

Oxford   New York

Athens   Auckland   Bangkok   Bogota   Buenos Aires Cape Town
Chennai Dar es Salaam   Delhi   Florence   Hong Kong   Istanbul Karachi
Kolkata Kuala Lumpur   Madrid   Melbourne Mexico City Mumbai Nairobi
Paris Sao Paulo Shanghai Singapore Taipei Tokyo Toronto Warsaw

with associated companies in Berlin Ibadan
Oxford is a registered trade mark of Oxford University Press
in the UK and in certain other countries

Published in India
By Oxford University Press, New Delhi

ISBN 019 5655338

Typeset by Jojy Philip, New Delhi 110 042
Printed in India by Roopak Printer, NOIDA, UP.
Published by Manzar Khan, Oxford University Press
YMCA Library Building, Jai Singh Road, New Delhi 110 001

In memory of
S.P. Singh
and for the ceaseless journey of
Jeremy Seabrook

Acknowledgements

For many years, my partner Charu Gupta nudged me to compile a volume of my scattered writings. To acknowledge her seems frivolous, for she is as much a part of the book as myself, despite the pressures of her doctoral research. This book would never have begun, let alone been completed, without her.

Mahesh Rangarajan, my friend since college days, sat with the articles and worked sensitively to bring the landscapes to life. He began a process, to which Ramachandra Guha gave further direction.

Richa Singh from Jawaharlal Nehru University kept me on my toes with constant inquiry of the book. Her lively company has been a treasure, and I hope I have lived up to her expectations

Nandini Gooptu and Barbara Hariss-White of the University of Oxford, made the actual writing of the book possible. During my visiting fellowship there, they gave me whole-hearted support. To an extent, I owe this book to both of them.

From the late 1980s, many have played an important role in my evolution as an environmental journalist. Foremost among them, Anil Agrawal, a pioneer in this field, sensitized me to many of these issues. Under the journalist fellowship scheme of the Centre for Science and Environment, I was able to cover some of the relatively unexplored environmental concerns. The *Navbharat Times* under the late Rajendra Mathur and S.P. Singh gave me the opportunity and the freedom to work on these issues. Without the *Economic and Political Weekly*, I would have felt slightly lost. Krishna Raj was always responsive to my reports, and gave them regular space in the journal.

I met N. Ram, Editor, *Frontline*, in 1990, when I went to receive the Statesman Rural Reporting Award in Calcutta. Since then, he has always been most encouraging, and this has helped me continue my pursuits. Mrinal Pande, author and journalist, is someone with whom

I could relate closely. I am also indebted to Darryl D'Monte for the initial impetus to my work.

Ann Ninan, Editor, *Inter Press Service*, initiated me into writing on environmental issues in English. In her unique way she is present in many of these reports. Kunda Dixit of *Panos* encouraged the idea of this book from the onset. J. John, with whom I edited the monthly journal, *Labour File*, made me conscious of coastal issues. Daniel Nelson of *Gemini News Service* took a keen interest in releasing some of these reports.

Jean Drèze, Jeremy Seabrook, M.S.S. Pandian, and Shekhar Krishnan have not only read through many of these pieces, but also helped in improving them. I am also grateful to Rajesh Tandon and Harsh Jaitli of PRIA, Arun Kumar Singh, Nabho Datta, Souparna Lahiri, D. Thankappan, Kanak Mani Dixit, Brij Mohan Gupta, Ramesh Pahari, Sunita Narain, Subodh Roy, Devchandra Jha, Gangesh Gunjan, Usha Rai, Bharat Dogra and Sunil Kumar Chowdhary who have contributed to the work.

Many friends have always been there when it mattered: Anand Swamy, Chandresh, Jai Gopal Goswami, Jai Lal, Aditya Nigam, Ajay Kumar, Bijoy Basant Patro, Sanjay Patro, Om Prakash Bhatt, Faisal Anurag, Rajendra Bhardwaj, Chatar Singh, Yadvendra Pandey, Gaurav Dutt, Animesh Shrivastav. A special thanks is due to Ram Kripal Singh, Editor *Navbharat Times* on whom I could rely for many things.

People like Thomas Kocherry, Deepak Bharti, Gurudas Dasgupta, Chandiprasad Bhatt, Har Govind and Ram Kishore, with their immense courage, dedication and knowledge, have been a source of inspiration over the last decade. Similarly, my two teachers—Radhakrishna Sahay and Shahid Amin—have contributed immensely in the making of this book.

My family—Saryu Prasad and Shashikala Sharma (my parents), Lalit Mohan and Damyanti Gupta (Charu's parents), Nutan and B.P. Singh, Ritu, Vishal, and Diksha—have been with me through vicissitudes and hardships. My son, Ishaan, who is seven years now, was many a time my only company during remote field work. Through these journeys we came closer.

In the late 1990s, I started working with Gyanendra Pandey at a time when my life was going through many upheavals, and he gave me strength to continue my endeavour.

Given my constraints with the English language, I needed friends who could skilfully and patiently edit the articles. Aradhana and Manas did precisely this and without their help this book would not have been the way it is. A warm thanks to the staff of Oxford University Press, New Delhi, for their cooperation.

Finally I thank Sephis for supporting my Ph.D., because of which I was able to complete this book. Ulbe Bosma and Ingrid Goedhard from Sephis, in particular, have always been sharing and enlivening.

Contents

✦

Introduction

Landscapes and Lives is about the unfolding of environmental issues and movements in rural India during the 1990s. It makes no claim to a comprehensive coverage of the field. The reports in this book appeared originally in a number of publications, including the *Economic and Political Weekly, Frontline, Down to Earth, Inter Press Service, Gemini News Service, Navbharat Times,* and *Udbhavana.* They are concerned with issues which received little or no attention in the mainstream media. The focus of the book is on how rural people, movements, organizations, governments, and non-government organizations are confronting, or evading, the question of the ecology of land-scape. It deals with some marginal to marginal ecologies like *diara, khadar,* and ravines, where from the mid-1980s to the mid-1990s alone, thousands of acres of land and hundreds of lives have been lost, leaving once-prosperous villages devastated and isolated, and several families near-destitute. Yet this crisis, a festering sore on the landscape, rarely figures in any environmental agenda because these areas have little political clout and, perhaps, because the decline here is not sudden, dramatic, or imposed from above. It is a slow, everyday phenomenon (and therefore, perhaps, more pervasive), which does not fit into a neat category of crisis perpetuated by the modern developmental models.

Landscapes represent specific natural and social environments. They bear the marks, not of a fixed relationship with humanity, but of changes brought about by the impact of shifting political, economic, and social actors upon them. In them we see the effects and the process of, say, land erosion, deforestation, desertification, or pollution. At the same time, we can grasp the political and economic pressures which generate, exacerbate, or reduce these problems. We can see how the majority of people—tribals, small-scale farmers, fishing communities, boatmen, and labourers—have struggled to retain control over their environment, often in struggle against the state or business and, as we can see from the examples, they are by no means always the vanquished in these conflicts.

Rural landscapes have both an everyday and an episodic aspect. The way people, society, state, and the market function on a day-to-day basis is responsible for pollution, deforestation, waterlogging; while floods, the shifting of riverbeds, landslides, which are massive occurrences over long time-periods, are episodic. But there is a relationship between the two and sometimes, as in the case of diara region of Bihar, the khadar region of Haryana, the ravines of Madhya Pradesh, and the coastal belt of Kerala, we see the episodic becoming the everyday. With floods, landslides, and erosion becoming almost regular occurrences, impoverishment and trauma have become everyday reality for many people. It is essentially the poor and the weak who bear the brunt of these disasters. These processes of ecological and economic marginalization are leading to extreme forms of criminalization, violence, and conflict.

The link between livelihood and landscape is most clearly visible in the rural areas, where a majority of people use not only land close to where they live, but also land far from their home; they depend on forests, rivers and coastal waters for domestic consumption or for resources they can sell. The strategy for survival based on natural resources by the rural poor is varied and complex. It is governed by the sheer need to survive, along with a profound knowledge of the locality and its resource potential. The poor find ways to circumvent, overtly or covertly, the formal and official economic arrangements. However, in the face of state power, they do so at the regular risk of being harassed, caught, fined, jailed, and even killed. The continuous killings of coastal fishing people in Gujarat, Maharashtra and Tamil Nadu, and police firings on tribals in Madhya Pradesh demonstrate the vulnerability of people, desperate to find resources for survival. In all probability, this risk will get worse, as the resource conflicts in the coming years intensify, as more people fight over less.

Environmental issues are often inextricably caught up with the local, social, or developmental problems of rural India. A village constructing its own dam, or a region struggling to reclaim the wasteland, or women in the Himalaya striving for basic facilities, together with their afforestation campaign, are not necessarily focusing on environmental issues. They are nevertheless engaging in some way with developmental issues. They are looking for plausible, concrete solutions to environmental dilemmas of whatever type, from whatever quarter. There is no bias against collective solutions. Given a reasonable development policy proposal, they generally accept it, or at least give it a hearing. It is because of this readiness that new government schemes can be counted on to improve environmental policy.

A majority of the conflicts in the 1990s have their origins in the issue of use and control over natural resources. Who owns the land? Who has access to the trees and produce of the forests? Who will cast a net into the seas? These questions have emerged in new forms, and even hitherto undisputed areas like ponds and ghats, ravines and wasteland, and deep seas have become a bone of contention between the state and business on the one hand and the rural poor on the other. To this scenario are now added new pressures on natural resources, exacerbated by economic liberalization and globalization. Some structural adjustment policies, large-scale infrastructural works such as highways, irrigation schemes, and hydroelectric power plants, and the increasingly footloose nature of polluting and hazardous industries have made a mockery of existing environmental regulations. The threat of appropriation and exhaustion of certain natural resources in the service of global capital and markets, have often set people against government. Indian deep seas and coastal areas have for many years been a virtual battleground, between those who want to maximize foreign exchange and those who depend on the ecosystem for their basic needs.

More than this, the question of resource use and conflict is not only a question of confrontation between local society and the state, but is also an aspect of the relations between certain neighbouring countries in South Asia. The collective plight of Indian, Pakistani, and Sri Lankan fishing people in border areas is an example of this.

There has been no such thing as a single perspective or a single movement on environmental issues in the 1990s. Given the vast range of environments, and the diversity of the people living symbiotically with and claiming rights over their landscape, there is a corresponding multitude of issues and perspectives, as well as of organizational initiatives over the environment. Differences appear within the countryside, between regions or occupational groups, between rich and poor, between upper and dalit castes, between men and women. Women in a village in Bihar may differ with men over the choice of plant species. Poor villagers of Haryana, Uttar Pradesh, and Bihar fight with each other over the use of land, which is formed by the silt deposits from the waters of the Ganga and Yamuna rivers. Dalits in Bihar or Orissa resist the appropriation and near-exclusive control of public water bodies by upper-caste landowners.

Similarly, many organizations, campaigns, movements, and alliances are constantly appearing in remote areas, addressing very specific local environmental issues. Each is fighting to stop or to build a particular dam, to resist displacement, to claim rights over land in forests or the fish of the seas, or to protest against the cutting of trees. Each struggle chooses its own strategy, depending upon the nature of the main issue, upon the socio-cultural life of the people, and the character of the local organization. Some of these movements, on the margin of rural India, take their cue from and seek to identify themselves with broader ideas and movements, which will strengthen the base of their cause; but for the most part, they remain fragmented and scattered. Those who protest against industrial pollution do not necessarily attach themselves to the struggle for ecologically sound land use. From a social and organizational viewpoint, being affected by one issue does not automatically lead to a willingness to take collective action with respect to a range of issues. Given the multiplicity of the conflicts and the particularity of each one, the common identification of the causes and united action across different issues look like a very distant possibility. This is compounded by the fact that many initiatives on environmental issues are broadly based on single agenda.

There have been inspiring successes in the struggle over natural resources. These were intimately linked to the question of entitlement and social justice for the poor, and were animated by the conviction that social justice is vital in resource use. However, advancing claims over resource use does not necessarily lead to the development of a sustainable system of exploiting such resources. The poor also tend to shape and structure their life in imitation of the dominant economic forces. Thus, a tribal village decides to cut the forests which were created by its own efforts. A farmer in a green revolution state seeks to drain out all the groundwater from his locality, and a coastal fishing community gets caught up in the compulsions of intensive fishing.

Some inspiring and persuasive leaders have emerged on the ground. Many have contributed considerably to the creation of a wider sensitivity towards environmental issues. Movements which they lead, whether anti-dam or against a new deep-sea fishing policy, have shown enormous skill and readiness to enter into dialogue with those in positions of political power. To some extent, they cut across class, gender, and regional divisions. They serve to radicalize social and economic concerns, because they make a direct link between the mainstream development model and the general deterioration of the natural environment. A wide cross-section of people—students, teachers, journalists, scientists, academics, poets, and writers—have taken an active interest in environmental movements. Numerous non-governmental organizations have also made their presence felt through issue-based campaigns and continuous constructive activities. At the same time, environmental movements have sought to associate themselves more closely with these NGOs.

The relationship between environment, labour, and labour organizations has also undergone a critical shift during the 1990s. A number of environmental activists and organizations have had recourse to the courts to challenge environmental destruction by industry. The closure or relocation of polluting industries has made thousands jobless, and labour and trade unions have resisted this. In spite of this, a positive relationship has been emerging—however slowly—between trade unions and movements for environmental protection. Following the initiative of an independent union of fish-workers, all the major trade unions in the country came together to protect the sea and fish stock. They sought to address issues of the management and sustaining of resources, marine ecology, technology, exports and markets, and consumption patterns and food security. Industries which poison the air and water also cause great harm to the health of workers and people living in the vicinity of polluting units. Hence, trade unions and labour support groups have looked for alternative ways and means to an enhanced version of freedom—freedom from oppression and from pollution. Again, at the initiative of an Indian trade union, people from several continents gathered in Delhi to form a 'World Forum of Fish Harvesters and Fish Workers', and showed that the era of globalization calls for a different kind of internationalism of labour in the struggle to protect both labour and the environment.

At a time when a combination of government and market is constantly and deliberately altering patterns of production and distribution, it becomes clear that efforts by communities all over the country, who are managing their resource base competently, call into doubt the economic dominance of both government and market. The Indian economy is seeking to absorb a number of independent working and living systems, which are not yet integrated into it. Such communities, which have developed in places all over the country, exhibit a sense of caring, collective action and self-governance, and responsibility for harmonious human/nature relations; and these have been sustained over time by traditional institutions, patterns of leadership and values. There are rajas and *sarkars, halmas* and shark hunters. They may be small scale and rendered invisible, but they manage to uphold, against all odds, an ecological version of living well. At the same time, it is hard to imagine how these homogeneous and hierarchical social arrangements can coexist with an increasing desire for social justice, and ethnic, religious, and cultural diversity in a country as vast as India. The raja here has little commitment to broadening, democratizing, and deepening the social and political life of the community, while the sarkar, on the basis of a single communitarian identity, goes along with the upsurge of the New Right.

Many examples stand out and catch our attention—for their land or water management, for their successful experiments in rural development, for self-governance, for their everyday beliefs, practices, and cultures. Some are rooted in a critique of modern industrial society and of the culture of greed and individual acquisition. Many are run by non-governmental organizations, which become involved in causes, as a response to concrete problems in their immediate neighbourhood. However, some of these 'green villages', which acknowledge a single authority or one organization, sometimes seek to regulate the actions and interests of their inhabitants through authoritarian commands, rules or laws, often backed by force. Enhancing the material welfare of the villagers by effective use of land or water resources can also be exploited to consolidate or expand the basis of moral authority. This also shows that sustainability in itself is not a sufficient condition to guide us in our search for democratic institutions and for meeting the manifold challenges in Indian society.

The role of the government in management of the environment provokes strong reactions. The Indian state commands enormous power in environmental matters. However, as a facilitator of the free market economy, the state has itself become the generator of new and previously unknown environmental problems. Government power of policy making, legislation, and enforcement is pervasive, but there is still room for organizations and popular movements to press for change. Faced with a growing number of grassroot movements, the government frequently prefers legislative or procedural methods, which have less to do with concern for the integrity of environment, and more to do with pragmatism. It has become a familiar phenomenon, that for people to establish a substantive right to control their environment, they must first go through a process of establishing procedural rights—for example, the right to information, the right to participate in government committees, and the right to appeal. The struggle for such rights within existing laws may strengthen the practice of Indian democracy.

The local environment has recently become a site of continuous struggle all over India, and will remain so. Even when the issues arise in response to a crisis, or as a campaign to stop some developmental project, or even to maintain status quo, they may be the precursors of broader forms of organization, equality and justice in society. However, the linkages between these two forms of consciousness are, through whatever urgent issues they express themselves, related to caste, gender, politics and the state. The organizations and movements that are articulating environmental issues, especially where these connections are made, are the organizations of the future, however small, local, or remote they may appear at present. At a personal level, they are not only clippings, they are clippings and me. At places, I am revisiting pieces of my memory—in Chotanagpur, in *diara*.

The Dispute over Water

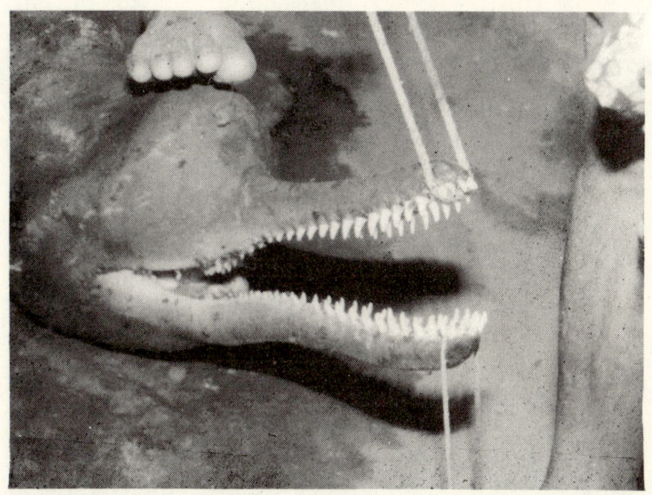

Susuk (Ganga dolphin), the only freshwater mammal on earth

Bateshwarthan Ghat, Naugachia district, Bihar

A Private Ganga

In this age of 'social justice', the ghats along the Ganga in Bihar are still under the control of zamindars and their contractors. However, a struggle to free the ghats, by boatmen, farmers, and fisherfolk, has now emerged. But this movement has been subjected to brutal suppression: boatmen and fisherfolk are being beaten up and killed, activists captured and dogs set on them. Boats have been attacked and sunk.

The zamindari system was legally abolished in 1954, but in Bihar it was actively practised till 1991. Zamindars had exclusive rights to fish, ply boats, and lease out demarcated areas of river to contractors and individual fish workers. An eighty-kilometre stretch of the Ganga in east Bihar (from Sultanganj to Pirpainti) was under the Jalkar zamindari or the Purvi Kshetra Marklaspur Jalkar zamindari, controlled by Musharraf Hussain of Murshidabad. Another stretch (from Barari to Sultanganj) to the west was under the Paschimi Kshetra Chanan Jalkar zamindari controlled by Mahashey Ghosh. This system of control was first challenged by the movement of fisherfolk, farmers and boatmen, under the banner of Ganga Mukti Andolan (GMA). After a tremendous struggle, the GMA achieved a major victory in 1991, when the Bihar government abolished Jalkar zamindari or *panidari*. However, this did not put an end to zamindari control of the ghats.

The Ganga and its ghats dominate the river system in Bihar. The river touches the Bihar–Uttar Pradesh boundary near its confluence with the Karmanasa, and inside the state, the Gogra, the Gandak, and the Son join it at places not far from Patna. Further east, the Punpun joins the Ganga from the south at Fatwah. The other tributaries are the Harohar and the Kiul, which join the Ganga at Surajgarh in Monghyr. The Gongra, the Gandak, the Burhi Gandak, the Kosi, the

Published in *Frontline*, 19 April 1996.

Mahananda, and their tributaries join the Ganga from the north; and the Karmanasa, the Son, the Punpun, the Phalgu, the Sakri, and the Kiul from the south.

The Ganga and its tributaries form the lifeline for millions of people, comprising fisherfolk, farmers, labourers, petty traders, and many others. The ghats are used either for fishing or as ferry points to commute daily to the nearby towns to sell products and earn a living. The ghats, which form an essential part of human living, are a part of the diara region spread over 9 lakh hectares. A large part of the diara land lies inside the riverbed and is used for cultivation in summer. It gets flooded in the rainy season. The diara land on river-banks is available for cultivation for eight to nine months in a year. Above these are upland diara, inhabited by millions. The ghats are the only link between this region and the outside world.

Officially, hundreds of ghats in Bihar were public property but in practice the zamindars, the contractors, the landlords and the criminals, working in close nexus, treated the ghats as their private property. A Bihar government gazette notification of August 1978, listing 108 ghats in Darbhanga district, 42 in Muzaffarpur, 17 in Saran, 23 in Champaran, 62 in Monghyr, 52 in Purnea, 19 in Santhal Parganas, 13 in Singhbhum, and 18 in Palamu, does not take into account all the districts in the state. The ghats were governed by the antiquated Bengal Ferries Act of 1885, which was in force only in Bihar. Under this, the Bihar government auctioned and leased out the ghats to private contractors and zamindars, who collected tolls and taxes for every activity related to the ghats. They also plied ferries or boats and sometimes even hired boats. No other person was allowed to ply boats or make any other use of the ghats in any way.

A payment had to be made to cross the river or to take vegetables, milk, cattle, grass, and so on, by ferry. Fisherfolk and boatmen were taxed on fishing and rowing boats. Poor people who crossed the river daily to sell vegetables and food-grains were grossly overcharged.

Ram Saran, a GMA leader from Bhagalpur said, 'It is unbelievable that the Jalkar zamindars and their families have been controlling the Ganga ghats in this region for the last hundred years. Though the Ganga flows free, it cannot be used freely by the common people'.

Ram Kishore, from a village in Shankarpur diara of Bhagalpur felt that the control over the ghats was enforced by means reminiscent of goonda raj. According to him, criminals hired by zamindars ruled the scene and charged exorbitantly for everything. When he once refused to pay and tried to organize the villagers against this practice, he was attacked, and his life was under serious threat. He had to leave his village and hide for months.

The fisherfolk and boatmen were terrified and quite often bore the brunt of the activities of zamindars. Kailash, a fisherman from Kahalgaon with a small boat narrated how they had to give money and fish regularly to the ghat owners. Otherwise, they could not pass through the ghats or even fish outside them. They could not protest. Anyone who did, was beaten up. Vakil Mandal, a boatman, was badly beaten up and left in a den of dogs when he refused to pay. Another boatman of Dildarpur, Sadhu Mandal, was also severely beaten up.

Along the Sultanganj–Pirpainti stretch, there are three main ghats: Bhagalpur, Bateshwarthan, and Sultanganj. The Bhagalpur ferry station has three chief ghats, at Adampur, Barari, and Mahadevpur, which cover a stretch of about 35 kilometres. They were controlled by zamindar Bhaggu Singh's family for over a century. Bhaggu Singh's grandson Ratnesh Narain Singh controls them now. He recalled how the ghat and ferry business was introduced by his great grandfather. They were veterans in this business and their family controlled ghats in other parts of the state as well. Nobody else had steamer vessels in this region and those who did not have vessels could not get the lease. Thus, they managed to gain control of these ghats. Ratnesh Singh and his lawyer claimed that under the existing law, the government fixed the toll. According to them, this toll is collected only for creating certain fixed assets and providing facilities to passengers. They have the right to charge an extra fare. In reality, the tolls and charges are collected arbitrarily and they are inherently exploitative. For example, at the Adampur ghat, passengers pay Rs 4 for a steamer ride, against the official rate of Rs 2.40. A basket of bananas attracts a charge of Rs 3, against the official rate of 36 paise.

The situation is similar at the Bateshwarthan ghat in Naugachia district, controlled by Laxman Singh for the past forty years. No steamers operate at this ghat. Hired hands ply Laxman Singh's boats, charging Rs 4 a passenger, of which Laxman Singh gives only Re 1 to the boatman. Boat owners are not allowed to ferry passengers. They can ferry certain goods but only after paying Laxman Singh, as well as certain criminals.

Much of the boatmen's income goes into the pockets of people like Laxman Singh. According to Upendranath Chaudhry, a GMA activist from Bateshwarthan, formal control of the ghat was given to Laxman Singh in 1993. Earlier, he controlled it without any legal sanction. Complaints to the district administration had no impact on his fiefdom.

Arun Kumar Singh, District Magistrate of Bhagalpur, explained that though technically there was no restriction on crossing the Ganga, certain portions of the river, each covering a stretch of 2–3 kilometres, had been demarcated at a number of places. The government controlled these public ghats technically. However, these were always auctioned to certain people. The administration fixed the minimum reserve amount according to the traffic and the tolls for a particular area and then auctioned it. It was the responsibility of the successful bidder to run steamers across it, construct jetties, arrange for the movement of cattle, goods, etc.

He identified two reasons for the continuing private control of ghats. First, the government has no resources to take care of them. Secondly, the Bengal Ferries Act invalidates any government effort to restrict or control the private parties or ex-zamindars. When, in response to some representations he exempted certain classes of people, especially the poor of Shankarpur diara, from paying the toll, the High Court declared the order *ultra vires*.

It is surely an anomaly that the ghats are still governed by the Bengal Ferries Act, which came into force on 1 August 1885. This Act was the basis for the *Bihar Ferries Manual* prepared in 1950 by the government of Bihar. An amendment made in 1977 empowered the state government to arrange for the control of ghats or ferries without any auction or public notice. Though the public and private ghats and the ferries are the exclusive property of the government under the Act, the government cannot cancel a long-term lease if a private party violates the agreement. The maximum punishment for any such violation is a fine of Rs 100. The Act was framed essentially to serve feudal or colonial interests. It referred to the zamindars and their people as 'ferry farmers', who were responsible for ferrying across 'the passengers carefully, safely and in the least possible time'. Its origins lay in the key role of landed interests in helping the sword-arm of the Raj.

Ferry farmers were not allowed to demand or charge tolls from 'His Majesty's Regular Forces, Imperial Service Troops, when on duty, or on the march, all members of the families of officers, soldiers or authorized followers of His Majesty's Regular Forces or any local corps, all carriages and horses belonging to His Majesty or employed in His Majesty's military service, all carriages and horses when moving under the orders of military authority'.

The *Bihar Ferries Manual* provided a list of items that invited freight charges: *Palki* (palanquin) with six *kahars* (sedans) loaded, coaches (*baggis*) drawn by a horse or by one or two oxen with *sawar* or groom, four-wheeled carriages drawn by two horses or a servant, horses or camels accompanied by a groom, elephants with *hawda*s and so on. The list also included essential commodities such as milk, vegetables, grass, salt, sugar, coal, coconut, tobacco, fish, oil, and bamboo.

In addition to tax exploitation, travel by ferries has also become risky, thanks to the rickety condition of the vessels. 'Accidents' occur every year, especially during the monsoon. On 9 August 1995, fifty persons, mostly women and children, were drowned in a boat capsize at Bakhtiyarpur, near Patna. After this, Chief Minister Laloo Prasad Yadav announced that the government would start a ferry service at the Sidhi ghat where the accident occurred. In 1993, forty persons drowned at the Singhkund ghat in Naugachia district. In 1998, a mishap at the Doriganj ghat in Chapra claimed seventy lives.

Most of the private steamers and ferries are unfit to carry passengers, a fact to which repeated attention has been drawn by numerous inspection reports of the Deputy Director of the Inland Water Transport Department and Chief Surveyor of the Inland Water Transport Authority, Bihar. Because leases cannot be cancelled even if contractors and zamindars violate the rules, it encourages the flouting of safety standards. Thus, the two steamers at Munger ghat do not follow even the minimum safety rules. They do not have life-jackets and lifeboats, hand pumps along with a power pump to take out accumulated water from the bottom of the vessel, fire extinguishers or sand buckets, and so on. The navigators often do not even possess driving licences and documents of vessels' ownership. Further, private contractors and zamindars carry passengers and goods in excess of the vessels' capacity.

The GMA, in its continuing struggle against zamindari, urged the government through meetings, rallies, and demonstrations, to free the Ganga ghats from the control of zamindars, contractors, and criminals. In the Bhagalpur, Sultanganj, and Pirpainti regions, the GMA intensified the struggle, with the people of Shankarpur diara, Charnia, Faridpur, Lodhipur, Laxmipur, and Nanhakar villages in the forefront of this struggle. Many were owners of small boats who were compelled to pay toll for ferrying vegetables, food-grains, and fruits. If they unloaded the goods anywhere away from the public ghats, they were beaten up. They now refused to pay this toll and used other points to unload their goods. In other ghats also farmers, fisherfolk, and boat-persons led by the GMA started refusing to pay the toll.

The zamindars, seeking to repress the struggle, had the agitators beaten up and detained, their boats sunk with the cargo, and their houses looted. Yogendra, a GMA activist, said, 'Our movement is facing a life-and-death situation. The movement against Ganga zamindari continued for ten years. Now we are again getting ready for another long and tortuous struggle. But we expect the state government, which claimed to stand by the poor and the downtrodden, to respond positively to the situation soon'.

❧

Snatching Food from a Tiger's Mouth
Dalits and Water Rights

Who owns the thousands of ponds and tanks in northern Bihar is a highly contested issue. In Madhubani, Saharsa, Darbhanga, and Supaul districts, an intense conflict has emerged over the ownership and fishing rights in government and traditional ponds, rivulets, and waterlogged areas. Mallahs, the traditional fisherfolk, the Musahars, and other sections of dalits living close to these water resources are struggling hard to claim their legal rights over them. Predictably, the response of the upper castes, landlords, and moneylenders, who have controlled the wetlands for long, has been violent.

Several of these ponds are on land owned by the state government. Officially, the right to fish in these ponds and in the rivulets and waterlogged areas belongs exclusively to the Mallahs and the fishermen's cooperative societies formed by them, or to the residents of the dalit villages near these water bodies. These water bodies are used not only for water harvesting, but also for fishing and other quasi-commercial and commercial activities. Whether located in towns or in villages, they are controlled and exploited by the rich and the powerful in a variety of ways. 'It is hitherto unknown and unheard of that Mallahs, Musahars and other poor people are working as wage labourers in what are supposedly their own ponds and tanks. We are challenging this. In one block alone, Jhanjharpur of Madhubani district, Mallahs and Musahars have taken control of thirty-two government ponds near their villages in the last one year. The ownership and use of over hundred ponds in this particular block are disputed. We need not only land reform, but water reform also in this region', explained Deepak Bharati of Lok Shakti Sangathan, a social organization and movement of dalits. The organization was founded

Published in *Frontline*, 16 July 1999.

in 1992 inspired by the ideals of the socialist legend Jayaprakash Narayan, and has been initiating struggles over a number of land and water-related issues.

Northern Bihar, especially the region locally called Mithilanchal, has an abundance of ponds, streams, rivers, and their tributaries. Madhubani district alone has more than 1500 ponds, small and large. Floodwaters collect in ponds and rivulets every year, making available an abundant fishing resource for most of the year. Also, the recurrent shift in the courses of the rivers Kosi, Kamala, and Bagmati converts tracts of land into a deep sheet of water abounding with fish. This region supplies fish to entire Bihar and even outside Bihar, to markets in West Bengal and Assam.

Mallahs, in particular, claimed ownership of the ponds and tanks in the region on the basis of age-old stories, folk songs, and traditional practices; the legal status proved otherwise since the early 1900s. It emerges from a case study of women in the inland fisheries sector, by Francis Sinha, K.A. Srinivasan, Rajiv Kumar Singh, and Viji Srinivasan, focussing on the ponds in Andhrathari block of Madhubani district, that the ponds in the area were in the possession of the Maharaja of Darbhanga before independence. However, myths of the traditional fisherfolk, the Mallahs, and from the Ramayana clearly indicate that they were the original owners. The Mallahs, somewhere in history, lost their kingdom to the maharajas. The maharaja made way for the zamindars after independence. After a few years of independence, gradually, about three-fourths of the ponds were transferred to the revenue department. In 1980, the Fish Farmers' Development Agency (FFDA) came into the field. Ponds in good condition were placed under the control of the block development office. Ill-maintained ponds were under the FFDA, to be developed and leased out to traditional fisherfolk. However, this did not happen.

For the present, the Madhubani district administration has made it clear that fishing rights for government ponds are leased. The leases are given to fishermen's cooperative societies, which allot ponds to their members. The ponds are to be allotted exclusively for the benefit of the Mallahs, the poor, and the downtrodden. There were twenty fishermen's cooperative societies in Madhubani district and only one could exist at the block level.

These societies being the sole inheritors, ran the show and, in fact, powerful and corrupt nexuses controlled them. As with other government-initiated cooperatives in the state, the organization and management of the fishermen's societies were in deep crisis. In

Andhrathari, there are about two hundred government ponds and nearly four hundred Mallah families, but the fishermen's cooperative society has only seventy-five members. The local rich and the land-owners largely control the ponds, even when they are officially leased to a Mallah or Musahar.

The history of ownership of ponds in the region is a contested one. There are many ways to appropriate a pond owned by the government. One is to have it leased out to some individual. The rich and powerful capture it by brute force. Landowners procure a lease, either themselves or on behalf of a relative in connivance with officials of the society and the government agency. But the preferred mode is lease in the name of a Mallah or Musahar who is heavily indebted to the landowner/moneylender. The landowner/moneylender controls everything, from the procurement of the lease to the sale of the produce. The Mallah becomes a worker who is remunerated one-fourth of the catch. Sometimes, a poor Mallah has no option, but to approach a landowner/moneylender, to meet the requirements of government revenue, seeds, food, medicine for fisheries and in the process, he becomes a wage labourer.

With the emergence of the Lok Shakti Sangathan, the dalit villages or *tolas* asserted their rights. Soharai Brahamotar is a village of Mal-lahs, Musahars and Kumhars in Lakhnaur block of Madhubani district. None of the 250 households possessed any land except their household plot. They survived on wage labour or fishing. The male members also migrated to Punjab, Haryana, and Delhi. Tiwari pond, a government pond measuring seven acres, adjoined the village. The villagers had leased this pond for many years from the cooperative. Then suddenly, in the late 1980s, a landowner with political clout obtained the lease to the pond in the name of an outsider, Ajij Mia. Ajij Mia landed up at the pond one day with a clutch of hired thugs and threatened the people against fishing in the pond. They complained to the administration, in vain.

Finally in March 1993, the women of the village took the initiative. An all-women meeting was organized by Somani Devi, Americi Devi, and Fulia. They took a decision to expel, if necessary by force, all outsiders from fishing or undertaking any other activity in the pond. Accordingly, Ajij Mia and his people were made to discontinue their activities in the pond. Tension prevailed in the village for days to-gether. The landowner came with his henchmen and threatened to eliminate everyone. They fired shots in the air while the labourers responded with abuse.

Unfortunately, in August 1993, floods in the region damaged the western embankment of Kamala river. The entire village was wiped out and the pond was also filled up. The villagers had to begin from scratch. After the houses had been rebuilt, the women of the village considered the issue of the pond. However, the menfolk were not ready for it, in fear of further retaliation. The landowner and his goons were continually threatening them with court cases, arrest, and attacks. 'It was like snatching the food from a tiger's mouth. We were really scared', confirmed Sushil and other villagers. It became so frightening that the villagers decided to boycott Somani Devi for her insistence on action, and even her husband and son went along with this decision.

Slowly, the villagers overcame their fear and in late 1994, they decided to approach the local administration to undertake the measurement and demarcation of the pond. When this was done, the villagers decided on a new course of action: in 1996–7, they worked together for forty days to dig the pond.

Now Sohrai village controlled Tiwari pond, managed by a committee of nine members consisting of both Mallahs and Musahars. A new set of rules was formulated: the committee would decide who would fish and how much would be fished in a season. Those fishing without its permission will be fined Rs 500. Only big fishes were to caught. Of every Rs 100 from the fishing, Rs 75 were set aside for a village fund while Rs 25 were for the fishermen. Individuals had to pay for any special requirement of fish, for marriages or festivals, etc.

In the pattern of fishing since the new system came into operation, the villagers did the seeding and in April–May 1998, they began fishing. Out of the money this yielded, they deposited Rs 5200 in the village fund. In August 1998, they collected Rs 25 each to buy seedlings from the outside market, so that the fish resources of the pond remained intact.

There are three other government ponds near Sohrai village, which were under the control of big landowners. Two of them—Angaragia and Sohrai—were also filled up, and the landowners are using the area for vegetable cropping. The third one—Durga—is controlled by a forward-caste landowner and its fish catch is exclusively being used for Bhagwati puja.

Sirpur Musahari is a village of Musahars, five kilometres northeast of Jhanjharpur town in Madhubani district. A big government pond—Sirpur—is situated right in the middle of the village. This was under the control of a brahmin landowner of a nearby village for many years. 'Even our children were debarred from entering the pond. It was so painful to see our children being beaten for doing a little fishing. When the Sangathan was formed in the village in early 1990s, we decided to claim our rights over the pond. First, in 1993–4, we asked the landowner's labourers who used to fish in the pond, to share the catch equally with the villagers. When they refused, we went to the local government officials— the District Magistrate (DM), the Block Development Officer (BDO), the Circle Officer (CO)—and found out that the landowner had no lease over the pond. Apparently he had bribed an official of the society and had been using the pond,' narrated Asharfi Sadai in the village.

Garvi Devi, in the same village, recalled the course of the struggle that followed after they decided to occupy the pond. They warned the landowner's people not to come to the pond. But this did not deter them from fishing. One day all the children and women of the village surrounded the pond and forcibly stopped them from entering. Some altercation took place, but they succeeded. Since 1994–5, Sirpur Musahari has owned Sirpur pond, but legally, the pond has still not been leased to them by the Fishermen's cooperative society. The villagers guarded the pond day and night. They had saved Rs 10,000 from the fishing. In 1998, they had a good harvest, so they saved another Rs 5000. They utilized this money to repair the school building and to purchase a petromax lantern. They also decided that the money saved in the village fund from fishing would be used for the care of the infirm and the elderly.

Only two kilometres ahead of Sirpur Musahari is Harbhanga village where a struggle over the ownership of a 2.5-acre pond continues. Hari Lal Sadai, a Musahar of the village, who took the lead in the struggle, recounted how in 1999 they had to face armed attacks from the landowners. In protest, they blocked the road and resorted to a dharna. The police intervened, but subsequently nothing happened to break the deadlock. In Haithiwadi village in Jhanjharpur block, since 1994, when they formed their own society, thirty-one ponds have been liberated in and around their village under the impetus and inspiration of dalit women.

All the Mallah and Musahar villages complained that there had been no support from the administration in their struggles. This is reflected in the fact that even after the liberation of the water bodies, most of them are still not being leased to dalits. In addition, hundreds of ponds in the region obviously continue to be under the control of landowners or moneylenders, but there is no official intervention.

In Prasad village of Madhepur block in Madhubani district, this issue could have taken a violent turn. Here a waterlogged area of 2700 acres, fed by a natural stream, was extremely rich in fish resources. A Koili tola of Mallahs and Musahars was dependent upon this area all year round. However, the big landowners of the village wanted to gain a stranglehold on the water body. In April–May 1998, several incidents took place at their behest. Surat Sadai and Mongia Devi of Koili tola gave an account of one of these incidents: 'Dozens of armed people from Prasad village arrived at a fishing spot. They damaged our small boats, looted nets and other equipment, and tried to compel the fishermen to abandon their work. Our people also gathered there, with whatever they could lay hands on, to counter the attack.' Many such events have been occurring at other places in recent times.

A region with a rich history of ponds, tanks, and other ways of water harvesting witnessed a cry for justice from dalits who wanted these water bodies to be restored to them. The Mallahs–Musahars and their organizations were not only willing to revive the ponds and streams, but also wanted to take possession of them.

Anxious Breaths in the Dark

Tubewell-related deaths in the green revolution districts of Haryana have been caused by the victims being trapped in pockets of poisonous gases. These are formed, it is said, by the heavy use of fertilizers. Since 1987, there have been hundreds of tubewell-related deaths in Haryana, in particular in the districts of Ambala, Karnal, Kurukshetra, Panipat, and Sonepat, where fertilizers are intensively used.

Some scientists assert that the gas-pocket problem was made acute by the steady fall in the water table because of excessive use of groundwater for irrigation. This resulted in increased concentration of pesticides and fertilizers in some tubewells. Others put the cause down to the gas that occurs naturally in deep wells.

In July–August 1988, tubewell-related deaths numbered 35 in the districts of Karnal, Kurukshetra, Ambala, and Yamunanagar, the greatest number being in Karnal. In many affected villages, rice transplantation came to a standstill as panic-stricken villagers refused to operate tubewells. In the period of 1990–5, at least 75 people died from poisonous gas alone.

Farm labourers, especially migrant, failed to report such incidents for fear of losing their jobs. Surjeet Singh of Nagala Megha village in Karnal district died when he was forced to re-enter his landlord's tubewell pit, although he had earlier complained of feeling suffocated. The landlord then threatened Surjeet's family and prevented them from reporting the accident at the block office.

Another troubling aspect is that of compensation. Some families were given compensation by the district administration in 1988, but none thereafter. Rajbala, widow of Sukhbir of Rasulpur village in Karnal district, said she received a mere Rs 9000 as compensation

This is an updated version of the paper originally published in *Down to Earth*, 31 October 1992.

from the government. Two brothers Balwinder and Lakhwinder died, leaving eleven members of the family to be looked after by a third brother Joginder, who also worked as a farm labourer. One widow bitterly complained: 'Is a man's life worth only Rs 9000? This is the compensation given for dead animals'. No medical care is provided for those who survive the gas pits and the precise effects of the gas often remain unknown. Labourers continue to live with physical injuries and scant medical attention.

Usually, farmers install electric motors for tubewells on the ground. However, over-exploitation by tubewells has drastically lowered the water table in the entire agricultural area of the state. This has forced the farmers to dig pits to install motors within an optimum suction lift, their depths reaching even 60 feet below ground level. According to a survey conducted by the Central Soil Salinity Research Institute (CSSRI) in Karnal, 60 per cent of the electric motors were lowered in 1988, 26 per cent in 1987 and 8 per cent in 1986. Furthermore, the pits were without pucca bricks and plastering and the pit floors were also usually unlined. Such pits have gained notoriety for being gas chambers and death traps.

Gas accumulated in these dry wells and deep pits. According to R. Chandra and N.T. Singh of CSSRI, the clinical reports of the victims confirmed death due to asphyxia due to lack of oxygen and excess of carbon dioxide. In the absence of oxygen, oxygenation of red blood corpuscles is prevented. The colour of the blood, nails, lips and body change to blue. There was congestion in the lungs and the brain, which pointed to carbon dioxide as the culprit. Scientists believe that carbon dioxide accumulates at the bottom of the pit. Decomposing organic matter such as straw, animal dung, and dry weeds present at the bottom of the pit, could also be a source of poison gas.

Whether or not it is carbon dioxide, the biological activities taking place in the surrounding soil are the major source of the poisonous gas. Soil with heavy use of chemical fertilizers contains a high amount of carbon dioxide. When it rains, the rainwater moves the carbon dioxide laterally to settle in a pit. A professor emeritus at the National Geophysical Research Institute at Hyderabad said that studies show that concentration of carbon dioxide is as much as 2 per cent of the air, whereas the normal concentration is only 0.03 per cent.

B.D. Dhawan, Professor at the Institute of Economic Growth, New Delhi, who studied groundwater issues, disagrees with this view. He doubts if increasing fertilizers and pesticides accumulating in well water is responsible for the accumulation of carbon dioxide.

According to him, the gas occurs naturally. But the difference is that now the water levels have dropped so drastically that farmers are placing their tubewell motors in deep pits.

The Groundwater Cell in the Agriculture Department of the government of Haryana has formulated a set of precautions to be followed before entering a tubewell pit. First, lower a burning lamp into the pit to ascertain the presence of carbon dioxide. Four to five kilograms of lime (calcium oxide) dissolved in water, should be sprinkled into the pit before someone enters it. Alternatively a table fan is to be lowered into the pit and run for at least an hour to force the carbon dioxide out. The fan is to be left on while the labourer works. Another suggestion is that the labourer should enter the pit with a rope tied around his waist, so that he can be immediately pulled in event of gas in the pit. In actuality, the only and temporary remedy seems to be that after an incident of tubewell-related death, people avoid entering the well or pit for some time. Landlords never really bother to take the necessary precautions, being only concerned with getting the job done.

The permanent remedies suggested are that the walls and bottom of the pit should be sealed with plaster. Wet wells with pits deep enough to touch the water table are also suggested because by exposing groundwater, a free water surface is available for absorbing gases. However, farmers find unplastered pits cheaper because the depletion of groundwater almost every year forces them to increase the depth of the well.

The Groundwater Cell has notified most of the agricultural areas in Karnal and Panipat districts as 'dark zones', where the authorities plan to discourage and ban new tubewells as well as subsidies, loans and power connections. The dark zone is one where the annual intake and use of groundwater is more than 85 per cent. A white zone denotes a 65 per cent annual intake of groundwater, and grey 65 to 85 per cent. All five blocks in Panipat district, four out of five blocks in Karnal district and six out of eight blocks in Kurukshetra district are in the dark category.

As long as the level of groundwater in the Green Revolution districts keeps falling, farmers will continue digging pits to draw out the maximum water from the ground. And Haryana's labourers will continue to face the agony of poison deaths in the gas chamber.

The Fickle River Basin

Yearly floods, diara village

Villagers coping with floods

Diara Diary

July 1992: heavy rains in Bhagalpur, Bihar have flooded its *diara* region. Thatched huts, household items, bhadai crops—all have been swept away. Many animals have drowned. The people of the diara take refuge under the shelters of plastic sheets they have erected on footpaths, roads, and Sendis compound. They stay here during the rainy season, and return to their villages in the diara after the fury of the monsoon is spent. For the inhabitants of the diara, this is a recurring annual feature.

The land between the Ganga and its tributaries—Burhi Gandak, Gandak, Kosi and Sone—is called the diara. Here the soil is mostly sandy and light. The diara is subjected to constant flooding and erosion. Low-lying diara areas, and those lying close to the currents of the Ganga and its tributaries, are flooded annually; while those located higher, and away from the current, are inundated sporadically. Erosion normally accompanies the receding floods. Due to continuous erosion and floods, and the changing course of the rivers, some areas emerge out of the water and some others get submerged. Some diaras are like islands between two rivers, or between two currents of the same river, but mostly they occur on both sides of the rivers.

A prominent marker of the Bhagalpur diara was the Bhagalpur Blindings Case, when many convicts were blinded in jails. Most of these belonged to the diara, and were locked up on charges of petty theft. Prior to this and even afterwards, the diara was a witness to many notorious massacres, like those of Shankarpur (1955), Bangalwa (1982), Medini Chowki (1983), Pipra (1984), Shambho (1985), Sonbarsa (1985), and Taufir (1985), in which an estimated 1500 people have been killed.

This collection of articles originally appeared in the *Navabharat Times*, 12–20 January 1990. This is a substantially revised version incorporating more recent developments.

In the villages, the diara is also called doab, which literally means a mesopotamia, or a tract between two rivers. Local villagers distinguish between two kinds of diaras: those which suffer the onslaught of floods and erosion once, or several times in a decade, and others, called *karar* diara, where erosion takes place after many decades or once in a hundred years. Among the several stories and myths connected to the diara, some can be traced back to the *Puranas*. According to a legend, the diara was an area under the curse of the king Bhagirath, being the place he had travelled to, to get rid of his ailment.

Diaras are spread all over Bihar, covering nearly 200–250 kilometres. They cover nearly 9 lakh hectares of the land: 2.40 lakh hectares in the Ganga diara, 2.30 lakh hectares in the Burhi Gandak, 1.40 lakh hectares in the Gandak, 1.50 lakh hectares in the Kosi, and 1.10 lakh hectares in the Sone. The inhabitants of the diara area are mainly of the middle and the backward castes, like Gangots, Mallahas, Dhanuks, Gwalas, Khairwars, Tantis, Bhumihars, and Rajputs. Herbert Risley identified the Gangots and Mallahas as the original inhabitants of the diara region. P. C. Roychoudhury's *Bihar District Gazetteers* (1962) also refers to Gangots, Mallahas, and Yadavs as the main inhabitants on the banks of the Ganga. The Bhumihars, Rajputs, Kurmis, and Koeris may not be among the original inhabitants of the diara, but they have always played a major role in the socio-economic structure of the area. Before independence, the Ganga diara land was mainly in the hands of an Agarwal family of Raj Baneili, Dilip Narain Singh, and Kamleshwari Prasad Singh of Monghyr and a trader family, Ram Gulam Sahu. Since independence, there has been little change in the socio-economic structure of the diara land. Some diaras have witnessed the rise of new landlord castes like the Yadavs and the Kurmis. In addition to the land, the landowners and a few rich people of the town today control most other means of production, like fisheries, milk, and vegetable production in the diara. Even the Ganga was under zamindari control in the area until the 1980s. Till date, there has been no official survey of the diara land, nor has there been any effort to properly measure it. Diara inhabitants increasingly complain that the regularity of floods and the extent of erosion have increased over the years, and hence the environment of the diara is worsening.

✢

It has become an annual phenomenon for the villagers in the diara region to be displaced. Life in the villages which suffer displacement and resettlement every year is full of uncertainty, insecurity, poverty, and disease. Ten to twelve feet below Bhagalpur town lies the Shankarpur diara, which was the worst affected by floods. There are eight villages in this area. Some are slowly eroding every year. Families which could leave the village, left it a long time ago, and only maintain their links with the place because of their agricultural land. The houses, made of *kash* grass, bamboo and straw, lie disordered. 'Since the last fifteen years, floods have occurred every year, destroying our village. There is a regular loss of life and property—houses fall, crops are swept away; animals are killed and diseases spread. Government assistance does not reach here', said Ramprasad of Shankarpur diara.

Villagers displaced by floods face a tough time for four to five months of the year. They take shelter in Bhagalpur town, where they find it difficult to get space to settle even in the open. 'Like animals we roam the roads and the streets, looking for shelter and food—at the railway platform or the bus stand, on the roadside or on government land. Almost every year, we have to go to the town, because we cannot go elsewhere. But the Bhagalpur administration has made no efforts for our temporary relief and rehabilitation', complained another villager, Jaikishan. The police extort money at such times. Lallan Prasad said that during the days of displacement, the police come frequently and arrest them on charges of theft. They have to settle illegally, which is why the police keep harassing them.

Food and employment become a grim problem in these months. The people can neither get work on the land, nor are there any jobs in the town. Kailash of Chauvanniya village found it very difficult to survive. Their stock of grains and money was exhausted within a month. Very few were left with enough money to hire a cart, set up a pavement shop or rear a cow. The wages for labour are also very low in the rainy season. Some ply rickshaws, and women work as maids in households. Loans even have to be taken from moneylenders.

Disease and death make life all the more difficult. In 1989, after the floods, Roshni Devi of Babu tola lost her two sons to diarrhoea. Harisharan of Chauvanniya died of snakebite. Choki of Shankarpur died of high fever. Cows and buffaloes also die of diseases, or lack of fodder. With their feeble means of livelihood also taken away, people barely manage to keep alive.

Erosion also displaces people. During the erosion of 1971, hundreds of families of Shankarpur diara were permanently displaced. Since their agricultural lands were in the diara, they could not settle far away. From 1971–85, they were driven from place to place, sometimes near the Shankar Talkies, sometimes to Buranath, or the railway station, or to Nathnagar. Ultimately in 1986, they were settled away from town, near the airport, in Jhurkhuria hamlet, a desolate, uneven place, where the government gave 40 bighas of land to settle about three hundred families. Because of bungling in land distribution more than half the displaced population did not get land titles. Many families, whose place of residence had not been eroded, obtained land titles through bribe. Also, members of the same family managed to get separate land entitlements. Many people from the town also got in, with the result that only fifty-two genuinely affected families could get the land.

Jhurkhuria has no water, electricity, nor a proper road. Many of the displaced have no means of employment. Their essential means of livelihood is petty shop-keeping and wage labour. Some have arable land in the diara. They sow the crop, but the fields are so far away that farming and tending the crop has become extremely difficult. Fodder shortage affects cattle rearing and, by and large, buying fodder from the town is beyond their means.

Another example of the neglect of the diara-displaced people is Company Bagh, the area of the town which touches the boundary of Bhagalpur University. A hundred displaced families of Shankarpur, Chauvanniya, Faradpur-Noordipur, and Ajmeripur came here in 1948 and their colony was declared legal in 1956. But there is nothing here which speaks of a town—no roads, drains, electricity, ration cards. A single hand pump supplies drinking water for the whole colony. Though they live in the town, for their livelihood they are still dependent on their old diara. They go there for farming, labour and cattle rearing. They are scattered here and there, without any fixed address.

✝

In the Taufir and Anthavan diara villages in the Kahalgaon block of Bhagalpur district, the majority of the villagers live in constant fear that the Ganga can consume their villages any day. Before the floods in 1990 the current of the Ganga was about four kilometres away, but now it is a mere half kilometre off. Some of the outer areas of the villages have already been lost to the river, and the erosion presents a frightful sight—an eroded quagmire of land, uprooted trees and seething water. The villagers know that their villages could disappear in one stroke, but they are forced to stay on, as there is no alternative.

When the villagers tried to find some land, the local administration did not approve of it.

Taufir, a village of nearly two hundred houses, had 1500 bighas of agricultural land before 1971. According to the villagers, so far 1200 bighas have gone under water. Anapur, Nandgola, Rani, Ikchari and other villages have met a similar fate. Since the floods of the 1990s, many more villages are on the verge of erosion. Villagers of the area stress that in 1992 the Ganga changed its course with great force and violence. There was increased flooding and erosion, and a greater loss of life and property. Even after the flood waters receded, the main current of the Ganga did not settle, and the threat of erosion remained.

Krishan Das, a venerable old man of Taufir, says, 'I have been living in this village since my grandfather's time. There have been many vicissitudes and many a time the land was eroded. We faced it all together. But this time, the erosion has left us stranded'. Middle-aged Harihar furnished more information about how they had sent their women and children to their relatives. The village school had closed down. Some people were living in the neighbouring villages. They came to the village by day, but went back before it became dark.

According to the village headman, they were prepared to settle anywhere. The only condition was that the land should be away from the Ganga and safe from erosion, so that displaced persons were not displaced again within a year or two. The administration had been repeatedly requested to either stop erosion or consider giving the villagers alternative land, but did nothing. Some years ago, when a similar erosion threat loomed over some prosperous diara villages of Sultanganj and Munger districts. The administration immediately arranged for huge boulders to form a *bund* on that end of the Ganga. The current stopped moving towards the village, and slowly changed course. There was extensive agricultural land in Nandgola which belonged to absentee landlords, where Taufir villagers used to work on wages. The villagers resolved in September 1990, to take over four bighas of this land and settle there. The sharecroppers also did not object to this. Some constructed their huts; most others had marked out their plots. The local administration then intervened on the complaint of the landlord, had the huts demolished, and arrested some people.

✢

The inhabitants of the diara die a thousand deaths. Since 1938 Ganga floods and erosion have annually affected the diara village of

Vikrampur, situated one kilometre south of the Bihpur railway station, in the Naugachiya block of Bhagalpur district. Everyone in the village has a tale of woe of the misfortunes that have befallen them over the last sixty years and more—how their houses collapsed, how the land was lost, how farming came to an end, how they ran from pillar to post for a livelihood, how for a year or more daily food had been a problem, how educating their children, or repairing destroyed homes, or getting out of the moneylender's net was impossible, and so on.

Vikrampur was originally situated about six kilometres south of its present location. There were about four hundred houses, spread over 30 bighas of land. After the erosion in 1964, no residential area was available to the villagers. Some distance away, there were fifteen bighas of land, belonging to landowners of the neighbouring village, some of which the displaced people had bought according to their means. In Harijan tola, about sixty houses are cramped in two bighas, each house getting not even one *kattha* to itself. There is no community or grazing land in the village. In the crowded village, open space of any kind is scarcely to be found.

Agricultural land in the village has eroded thrice: in 1937, 1975 and 1987. Since 1990 the village has been free from erosion, but most of the agricultural land is now under water on account of the annual floods. So much sand has spread over the land because of the floods that it cannot yield anything except *parval* (*Tricosanthes dioica*—its fruit is eaten as vegetable). Even when farming is done, it is capable of yielding only one crop, rabi.

Waterlogging is another problem. While constructing bridges and highways, proper provision was not made for water outlets. Where such provisions were made, there was inadequate maintenance. Flood waters remain for many months, harming the rabi crop. Some areas are permanently waterlogged. 'Almost one hundred acres of Vikrampur land has become permanently waterlogged. There are two reasons for this. First, there is no provision for water outlets. The second, is the vested interest of powerful people and criminal gangs. Such waterlogged areas have a lot of fish, and fish traders do not want the area evacuated until they have killed the entire fish population', says Devchandra Jha, who teaches in a local school.

No family in the village has more than ten bighas of land, and 10 per cent of the families, belonging to the Scheduled Castes, have no land at all. Of 450 families, except for the sixty brahmin families, all the rest—Kurmis, Koeris, Nai, Ravidas, and Muslim families—have been reduced to working as labourers to earn their livelihood. Brahmin menfolk have some employment in government or private services, though not of a kind that makes them secure and well-to-do.

Far from initiating any special project to stop erosion and provide water outlets, even routine government development programmes are absent in Vikrampur village. It has no electricity or drinking water supply. The school and the library buildings have collapsed. Sharecropping and wage labour having virtually disappeared; there are no disputes regarding labour or land. They were not involved in the fight over fish in the waterlogged area, too weak to resist the powerful fishing interests. Caught in their poverty, insecurity and helplessness, they live merely on the hope that 'Ganga *maiya* would give back their land some day'.

✛

The Ganga diara of Bihar has nearly three lakh hectares of arable land, which regularly gets layers of new soil and sand. The land closer to the river has a layer 30–60 cm thick, which gradually decreases in the land away from the river. Only vegetables can be grown on land with a thick layer of sand. The rest of the diara land is fertile because of the new soil left on it by the floods.

The diara arable comprises upland, medium land and low land. The upland gets flooded only during very heavy floods, or during August–September. Given irrigation facilities it can be cultivated through the year. The diara has very little upland. The medium land remains submerged in floodwaters for a maximum of three months, from July to October. The low land remain flooded longer, and can be cultivated only for four to five months. The land which emerges out of water from time to time takes many years to become cultivable. Two kinds of wild grasses—*kas* and *jhauva*— with their roots help make the land firm and heavy.

Floods quite often destroy the kharif crop, grown from June–July to September–October. Even and rabi crop, grown from October–November to March–April is sometimes delayed because of floodwaters. Drought compounds the problem further. According to one estimate only 3 per cent of the diara has irrigation facilities.

Villagers estimate that only 15 per cent of diara land is cultivable. Primarily, vegetables are grown for sale in the town market. A trip to the town market entails trudging over miles of sand, through accumulated water and slush, and crossing the river by boat. The villagers, therefore, prefer to sell their entire produce to the local merchant.

From the Shankarpur diara, cow and buffalo milk is sold on a large scale at Bhagalpur and other adjoining towns. For the cattle rearers a major problem are floods, which carry away cattle and make the area disease-prone. In addition, obtaining fodder is difficult during and after the months of flood. Unable to buy fodder from the town, the milkmen are often forced to sell off their cattle.

Two types of milk trade are found in Bhagalpur town. Land-owners, who are also prosperous cattle owners, hire local labourers and small farmers to buy milk from all around the village. Their work starts in the early morning and ends late in the evening, and they are paid daily. On a smaller scale, is the milk trade by *daihars* independently. They do not own cattle but buy milk from the villagers, paying cash with an advance taken from the town shopkeeper. Losses associated with milk business like split milk can set them back considerably.

In the Ganga diara, from Sultanganj to Peripainti in Bhagalpur district, there are about 40,000 fishermen and about 15,000 fishermen in the Kahalgaon diara. The general pattern of the fish trade here is that prosperous people in villages take *pattas* for fishery rights at particular places along the diara. They generally get their work done through contractors, who in turn get in touch with local fishermen. Different agreements operate in different areas—at some places they get 10 per cent of the catch, at others 25 per cent, at yet others they get a monthly payment, while in some cases they work in a kind of bondage.

When agricultural land goes under water, the owner of the land is not entititled to the fish in the water on his land. He may not even be allowed to come near the newly formed pond. The powerful in the village or criminal gangs grab it. Every year many lives are lost in the fights between landlords and criminal gangs over the control of these ponds. From Sultanganj to Pirpainti in the Ganga diara, dozens of well-armed criminal gangs operate, which also have a stake in fishery. Even after making the patta payment, local fishermen have to pay extortion money to these criminal gangs. The Superintendent of Police in Bhagalpur gave an account of this for 1987: On 24 May, the Nandana Mandal gang kidnapped a fisherman, Durga Sahni. On 3 June, Rajkumar Sahni was kidnapped. In September, eleven fishermen were kidnapped and later released on payment of ransom. On 14 October, the Hulo Yadav gang looted the catch of eight fishermen. On

15 October, a fisherman, Radhey Sahni was abducted and killed and his boat stolen. On 16–17 November, at least twelve fishermen were killed near the diara ghat in Pirpainti police station. A contractor, Buddu Sau, contracted these fishermen from Kahalgaon. In Rani diara, a rival contractor Bhola Mandal arrested them. This led to a clash of two criminal gangs, leading to the deaths. On 1 December, criminals injured Ram Lal while he was fishing.

Bihar's diara is mostly unsurveyed. Even where there have been surveys, either before or after independence, new land-related disputes have cropped up because of faulty surveys. The changing course of the river after floods, the land erosion, and the appearance of new land compounds this problem.

An example is the Dhamdaha area in the western part of Purnea district, which was badly affected by the Kosi floods. Between 1901–11, the Kosi current moved westward towards Bhagalpur, thus leaving the area flood and erosion free while also throwing up new agricultural land. According to a survey of forty villages in 75 square miles, in 1923–6 (*Final report on the survey and settlement operation of the Kosi diara area in the district of Purnea, 1923–26*, Government of Bihar and Orissa, 1927), there were many changes in the landownership pattern because of the change in the course of the Kosi. First, the old tenants occupied their own land. Since the land boundaries had been wiped out, the stronger tenants encroached on the land of their weaker neighbours. The Santhal and Gangot sharecroppers were removed from the land after three to four years. People from Chapra and Muzaffarpur came and occupied the newly emerged land. The Darbhanga Maharaja gave them land rights. Santhals and Musahars normally tilled this land. The Maharaja also arranged land for some co-sharers. These co-sharers paid rent to the Maharaja and got their tilling done by farmers who came from Munger, Bhagalpur and Santhal Parganas. Farmers who worked under the co-sharers were called *kolatdar*s.

In 1922, the collector of the area acknowledged that the entire area was teeming with land disputes and consequent criminal activities. A survey of the area was necessary, particularly that under the Darbhanga Maharaja (*Bihar and Orissa Revenue Proceedings*, No 11324, 1926). In 1923, the Revenue Department started a survey settlement of the Kosi diara in Purnea. The *Final report on the Kosi diara survey and settlement operation in the district of Bhagalpur and Purnea*, 1926 says that during the survey, powerful people of the area got the tillers and the local poor dispossessed, and got a legal stamp over their seizures. The survey officer remarked, 'After a tour of the area it is difficult for any officer to believe that the poor people of the village have merely some dismal residential land in some corner of the village and the entire agricultural land of the village belongs to people of some different district'.

A second survey in 1950, intended to establish the right of the land-owners on their land, was faulty from the beginning, being essentially based on the 1923 survey settlement. The new survey settlement once again dispossessed the poor, the landless and the sharecroppers. With this land grabbing began a new wave of land disputes and land struggles. In the Kosi diara and other areas of Purnea district, the pattern of landownership is as follows: Kursela family 12,000 acres, Ramgulam Sahu 20,000 acres, Maulchand 10,000 acres, Jagdish Chaudhary 10,000 acres, Vaksh Chaudhary 12,000 acres, Prithvi Chand 10,000 acres, Saryu Mishra 4500 acres and Moihur Rehman 4000 acres.

The survey settlement of Bhagalpur district was done in 1902–10 under P.W. Murphy, ICS, but the diara was not included in this survey. There was a special Ganga diara area survey during 1908–15, but its aim was only to fix the boundary of the diara area. The diara area was under Raja Todarmal, and different landlords were given land rights. The tillers were poor Gangots, but the ownership of the land was with the landlords and their leaseholders. This Ganga diara area was also disturbed because of the dispossessing of the small farmers. However, no effort was made to reduce tensions or to have a fresh survey in order to resolve the land disputes.

The survey settlement of Bhagalpur district, which commenced in 1950, ignored the diara. A survey settlement of the area was, however, ordered in 1958, but only after land disputes in the diara became extremely violent during 1950–5. The survey started in 1959 and lasted until March 1965. In this survey, the first in a hundred years in the area, farmers and sharecroppers were dispossessed on a large scale, as had happened during other surveys. In many areas, owing to the influence of kisan sabhas and other organizations of farmers and sharecroppers, records were made and maps prepared in favour of the land tillers. The illegal occupations of big landlords were also brought to light. Nevertheless, as soon as the survey settlement was completed, the Bihar government scrapped the survey (Memo no. 10241, 2–3 December 1965), declaring that the survey settlement had no 'permanent authoritative significance.' No official explanation was given for this decision nor was a fresh survey ordered. Similar is the plight of the diara in other districts of the state. The Bihar government's *Estate Manual* 1953 provides for the annual survey of the diaras, and instructions have been provided for implementation (viz., the land thrown up by the Ganga will first belong to the tenant who owned it before it went into water; if there is no previous owner, the land will rest with the government, etc.), but in the absence of authoritative and genuine land survey settlements, these provisions and instructions are meaningless. The result is dispossession of small farmers and sharecroppers, the complete control by landowners, the terror of criminals and official connivance in all this. In Mahadevpur mauja in the Shankarpur diara, for example, the district administration had apparently made a map of the area in the early 1900s. Thereafter the entire area suffered erosion and the land remained submerged in the Ganga for years. When this land emerged out of the water in the 1980s, landowners Nathuni Singh of Shahkund and Bhurimal Marwari of Naugachiya seized about 1000 bighas. In the fight between the old small farmers and the new land grabbers about fifty lives have been lost.

The land grabbers started making use of criminal gangs, in return allowing these gangs to cultivate hundreds of acres of illegally occupied land. Abhay Kumar Sinha captured about 500 bighas of land in Shankarpur diara in 1990 and distributed it to a criminal gang for sharecropping. Before that, he had hundreds of acres of crop destroyed and then had the sharecroppers dispossessed. Similarly, in Ubari Pat village of Chauvanniya diara, the crops and granaries of sharecroppers were looted.

✢

Many in Bhagalpur, Munger, Naughachiya, Kahalgaon, Purnea, and Katihar say that over the years, the flood havoc has been increasing. The floods reach the diara earlier than before, and the entire *bhadai* crop, which is reaped in August, gets destroyed. The floodwaters also stay longer, delaying the rabi crop. The force of the floods has accelerated, and the loss of life and property has increased. The pace and extent of erosion has also increased. Earlier, villages on higher land were safe from erosion.

It is believed that the floods bring new soil and chemical elements, which raise the productivity of such land. This also reduces the cost of fertilizers and irrigation. For this reason diara farming had been called the cheapest. Now villagers complain that production in this area has reduced. Ram Sevak Srivastava, an agriculture scientist in Sabour observed, 'New kinds of diseases have been plaguing crops and animals. The chemical nature of the soil and water brought by the Ganga has also changed, because of which the land is not becoming fertile, but sterile'.

The difficulties of the people in diara may be traced to the pollution of the Ganga, the felling of trees, and the wrong planning of bunds and embankments. A study by two Bhagalpur University professors, K.S. Bilgrami and J.S. Dattamunshi, showed that in the 256 kilometres between Barauni and Farakka, the Ganga is the most polluted near the Mokamah bridge. Here, industries like the Bata Shoe Factory, McDowell Distillery, the oil refinery, the thermal power station and the chemical fertilizer factories discharge their effluents into the Ganga. Bata and McDowell discharge 250,000 litres of used and dirty water in the Ganga every day. According to Bilgrami, in experiments done to gauge the poisonous nature of this part of the river, fish introduced in the Bata area died in forty-eight hours, and in the McDowell area water, within five hours. The effluvium from the oil refinery is so much that once the river caught fire near Munger.

Not only is the Ganga getting polluted, but its rate of siltation has also increased. From Patna to Farakka, the strong current of the Ganga in the diara carried a great amount of eroding mud. This mud piles up in the Ganga at different places. The riverbed is rising due to industrial wastes as well.

Diara forests have also depleted on a large scale. Vikrampur had an abundance of rosewood trees, which are hardly visible now. Kas and jhauva grass is destroyed in the competition to quickly grab the land. Landowners cut this grass and tried to cultivate the land. Commercial exploitation of these has also begun.

The several bridges, embankments, pumping stations and irrigation projects erected on the Ganga adversely affect the water flow in diara. A few days of heavy rain in the diara are enough to accumulate water everywhere. Many large diara areas of Rupauli and Kadava sectors in Purnea and Katihar districts have become permanently waterlogged. In the Naugachiya area of Bhagalpur district, flood prevention embankments have been built for roads, railway and the canals without much thought given to water exit mechanisms. The result is that diara areas prone to submergence face more intense submersion. When the embankments are breached, the people of the diara lose their houses.

The region today has a disturbed existence, the constant dispersion making people more inclined to crime, murder and litigation. Caught in an increasing river wrap, the region is breaking under the weight of water. Every area shows signs of instability, every village is haunted by the fear of imminent non-existence.

What We Did and What the Khadar Did Next

Wherever the floodwaters of the Yamuna reach, the villagers in Haryana call it *khadar*. A khadar consists of flood plains with flat valleys, merging with wastelands; flooding by channel flow; mixed vegetation with grazing cattle; sandy soil; and village households dispersed in the drier niches. Officially, in Ambala, Kurukshetra, Karnal, Jind, Sonepat, Rohtak, Bhiwani, Mahendragarh, Hissar and Gurgaon districts, the Yamuna river region covers 21,265 square kilometres, of which 79.5 per cent is agricultural land, 18.1 per cent non-agricultural land, 3.6 per cent residential land, and 2.4 per cent forest land. Peer Badholi village in Karnal district clearly shows some of the features which have now become almost a regular feature of the khadar eco-system. The flood of 1991 devastated the village and about five hundred acres of agricultural land was destroyed. Two hundred and six families have settled in the dry areas, at least four kilometres away from the previous settlement.

Flood and erosion have been a feature of Peer Badholi. The years 1914, 1934, 1966, and 1978 witnessed the river changing course, banks overflowing and land submerging. However, since 1978, whatever remained of the village and its land had to face the wrath of flood and erosion in quick succession. Each time the village silently lost some part of itself and then came 1991, when it faced the final assault.

After the floods of 1991, each family got from the government, as compensation for the displacement, a 25 square yard plot of land, Rs 400, two blankets and a loan of Rs 4000–5000 to build a house. Kartar Singh, who earlier owned more than eight acres of land, is idle. There is no work available in the vicinity, as floods and erosion have affected everyone else too.

The neighbouring villages of Lalupura, Sadarpur, Vada Daha, Devipur, and others also lost their houses, property, cattle, and crop in the floods, but most of their agricultural land was saved from erosion. They were not sure how long they would have to wait for an opportunity to plant the new crop. Mundi Ghari, a village of 257 houses, was submerged in flood water 4–5 feet deep in 1988. The submerged land was of no use for a year. Even today, signs of the rage of the Yamuna remain—deep excavations, wastelands with high levels of sand, withered trees, and damaged pathways. Mundi Ghari, which shifted to its present site in 1970, once had around 12,000 bighas of agricultural land, but 7,000 bighas have gone under the Yamuna. Some land has emerged from the water, but the thick layer of sand makes it unfit for cultivation. Some of those who have land want to sell it and settle down in some safer place. Those who had none left have gone as far as Panipat and Gharonda to labour. With less land in the village, less work is available, and the wages have declined to Rs 10 or 15 a day. Sharecropping, once prevalent, is not found anymore.

The three villages of Rasulpur, Mimarpur, and Garhi form one panchayat in Sonepat district. Only 16 kilometres away from Sonepat, Mimarpur village has seen the Yamuna coming closer and closer since 1980. 'We are losing our agricultural land and as we lose it, villagers in Uttar Pradesh are getting it. If we lose here, we should surely gain there', argued Sishpal. He demanded police protection to farm the land which had emerged from the Yamuna in the neighbouring state. Jainpur, Tikola and Nayabans villages have also lost their land to Uttar Pradesh and would like to forcibly claim the land somewhere else, at an appropriate time.

The khadar tracts of Faridabad district, especially Sholda, Bagpur Kurd, and Pahruka villages were seething with violence over the land question. Defying state boundaries, and injuries from firing by the Uttar Pradesh police, they thought it was their right to go to the Uttar Pradesh side, and farm the land in the Yamuna region. As the Yamuna slowly moved to the side of Bagpur Kurd, it was eating the land here, but was disgorging it on the other side. 'It is a question of life and death of 60–70 brahmin families. At stake is no less than 1000 bighas of land. Whatever comes, we are not going to leave it', was Balbir's logic. Tempers ran high here in the aftermath of the firing by the Uttar Pradesh police, and the subsequent injuries and arrests of a few villagers.

The agricultural land in khadar region is mostly sandy and not very productive. In many khadar regions it is called *kallar*, meaning wasteland. It is substantially deficient in the organic elements, especially nitrogen, zinc, and iron, being generally less than 1 per cent. Further, the crops are uncertain because of rains and floods. The khadar being a fragile and constantly changing region, the people here were more realistic and judicious in their agricultural practices in the past. The *gochani* tradition used to be practised here, which entailed mixed crops with less expenses, less water and modest productivity. The farmers' stakes were never very high. Livestock and the raising of animals were as important as agriculture. Mixed cropping, especially lentil and others, also helped to reduce the deficiency of the organic elements in the land.

The villagers are increasingly engaged in shaping the khadar landform, creating a new landform, modifying the existing one, or trying to erase it from the environment. They have substituted traditional crops with new high-yielding varieties, constructed embankments for regulating or diverting a channel flow, built dams or reservoirs upstream, and engaged in deforestation or afforestation of the local or the upper catchment of channels.

Earlier, the main crops were wheat, maize, millet, chilly and *arhar* pulse (*Cajamus indicus*), which required watering only once. Also, some of them like maize and arhar provided enough fuel. Over the last ten to fifteen years, the land is never left fallow. Rice has been introduced as a new crop, and the farmers alternate between rice and wheat, which has increased the consumption of water and fertilizer many times over.

Peer Badholi village in the khadar region of Karnal district mainly used to cultivate wheat, cumin, watermelon, sweet melon and some vegetables. Kartar Singh, from a farming family of many generations, used to harvest one crop a year some years ago, but now he tries at least two crops in a normal year. 'There is too much indebtedness. We want to produce more, so we have to invest more. It is like a circle', he explained. Pubnera, another khadar village of Sikh inhabitants, used to be heavily involved in dairying. Now this activity has considerably reduced. There is hardly any grazing land left within or outside the village, and the cropping pattern has consequently led to a decrease in fodder.

Not only has the agricultural life been mostly reduced to growing and harvesting rice and wheat, present agricultural practices have transformed the khadar region. High-yielding varieties, chemical fertilizers, pesticides, and tubewell irrigation have become the mainstay of agriculture with the advent of the green revolution. Whatever be the cost-benefit analysis of this, it is certain that this agriculture has further eroded the organic base of land here, more than in any other green revolution region of the state. This is a new external disturbance in the normal khadar eco-system.

Farmers in the khadar region experience all this with a deep-seated pessimism: 'The land was *pauli* (soft) before. It has now become *kardi* (hard). To get it ready for farming requires more labour. Even then, it is impossible to prepare it properly. It demands too much water' said a farmer in Mahiuddinpur village of Karnal district. 'We need excessive fertilizers. If we do not do that, the output decreases. If we reduce the quantity of the fertilizers, the produce declines', said Rameshchand in Datoli village of Sonepat district. 'Khadar has turned into a dry jungle. The land does not get an opportunity to breathe. The cost of agriculture increases year by year, but the output does not increase', said Niddi in Sholda village of Faridabad district.

Agricultural scientists are worried. A senior officer in the Agriculture department at Karnal, speaking on condition of anonymity said that the decline of organic elements in the khadar land is alarming. Nitrogen, zinc, sulphur, and iron are going down because of the new seeds and chemical fertilizers. Farmers use fertilizers in excess, especially nitrogen, to balance the deficiency, whereas there should be a judicious proportion maintained in the use of nitrogen, phosphorous and potash fertilizers.

This new agricultural system operates at more than one level. The carrying capacity of land having drastically diminished, in its degraded form it easily succumbs to erosion. At another level, the system exceeds the farmer's economic carrying capacity and makes an already weak khadar farmer more vulnerable. As the economics of the green revolution agriculture has become more or less standardized, and the floods more frequent, farmers are losing more of their crops and are doomed to indebtedness and poverty. Agriculture annihilates the khadar environment, leaving an alienated khadar inhabitant to create more khadar.

Kare, Sohan Singh, and Charan Singh, all from the Bahrampur

village in Karnal district, have similar complaints. Though they did the ploughing and sowing every year, they could not harvest the crop each time, leading to a heavy burden of debt. They already suffered this burden, with or without the flood. Deep Singh, a 60-year old *numberdar* of Mauja Bhud village in Karnal district said: 'I have 50 bighas of land. I took a loan of Rs 6000 for a tubewell ten years ago that I have still not been able to repay. Besides, we have to take loans every year for seed, fertilizer or pesticide. If the crop is damaged, we cannot repay the loan, but must keep paying the interest.' On account of the recurrent rains and floods, the debts keep recurring. Hari Singh of Peer Badholi, who has exhausted all sources of loans—money-lender, bank or society—in the last few years, is unable to repay them fully, and always feels the pressure of increasing interest. Kripal Singh of the same village, whose land was submerged by the Yamuna river recently, received a letter from the bank to repay the loan, or face the consequences of arrest and confiscation of property.

✛

The Yamuna is unable to meet even the day-to-day needs of drinking and irrigation water in the khadar region. All the wells and ponds of the region have dried out over the years. Getting water which is enough only for a few months, the region is now being officially declared a dry zone, in terms of groundwater availability. In Jebabad Kheldi village in Karnal district, in the summer months villagers have to trudge four kilometres to fetch water. Water for cattle has become a serious problem. This is in spite of the fact that all the forty families of the village have one, two or three tubewells. All the four old wells in the village are non-functional now. The situation is similar in Rasulpur, Chundipur, and Mahiuddinpur villages.

In Nagal Kala, a khadar village of Sonepat district, Kare Singh explained that for a population of 8000 in the village, not a single well is left for drinking water. All the six ponds there are slushy. Datoli, another khadar village, filled up all the three ponds and four wells in the village. Tubewells and pump-sets replaced wells and ponds in most of the khadar villages. In the khadar villages of Faridabad district, each family has a hand-pump for drinking water. One hand-pump costs Rs 1100 to 1400. In the summer months, they stop func-tioning unless they are extended deeper in the ground. Jebabad, Rasulpur, and Chundipur villages are going through a water crisis, in spite of having dozens of hand-pumps.

In the khadar villages of Karnal district, every three to four acres has a tubewell. Farmers in Shekhpur and Sholda villages say that a tubewell with three horsepower motor does not work now and they have to get one of five horsepower. Farmers in Rasulpur, Chundipur and Mahiuddinpur villages said that they now need tubewells of ten horsepower. On dry, sandy land when groundwater level goes down, tubewells have to be constantly dug up and driven deeper. They are lost in floods and erosion. They are burnt in times of erratic power supply.

Also, the new agricultural regime needs regular and copious quantities of water. 'Water does not rest here. Within twelve hours of watering, the land is dry again, and we can easily walk over it. However much water we pour in, it is absorbed', said a farmer, Nanakgiri Goswami, at Manoli khadar village of Sonepat district. Another farmer in the same village thought differently: 'To enhance productivity, we use more fertilizers, which demands more water for the crop'.

The end result is that many khadar villages have been declared 'dry' now. The Groundwater Cell, Haryana Agriculture Department, gave a long list of dry villages in the Yamuna region of the state, and outlined the measures being taken to improve the situation. In the dry zones, tubewells are not allowed to function and no new tubewells can be installed. But the law is only on paper. Some ten to twelve years ago, water was available at a depth of ten to twelve feet. Now it is 40 to 80 feet. To meet this lower level, the motor is now being installed in the unlined well which has been dug deeper especially for this purpose.

A few who can afford it, have gone even further. A farmer of Bakali, a khadar village in Kurukshetra district, has had a 30 horsepower motor installed, with a pipe of at least ten inches diameter, which goes 300 feet below the ground. It is called a 'Submersive' tubewell and sucks all the groundwater available in a wide region. The owner of this huge outfit exuded some confidence: 'I am trouble-free for at least ten years with this tubewell. It will only be dry when the whole district goes dry. Let the water go down, the pipe will catch it.'

This exploitation of water has become an important factor in the changing khadar eco-system. The increasing dryness, degradation, poverty, and impoverishment in the region is attributed to this. Therefore, the more it spreads over a longer period, the more serious is the danger. An agricultural scientist in Karnal has warned that with the sandy land and the lower groundwater level in the khadar region, irrigation water also has a tendency to go towards the groundwater level, taking sand with it. After ten to fifteen years, this creates a hard crust of sand and soil one or two feet deep on the land. Even a pickaxe cannot break it, nor can it be reclaimed.

✛

Land, being the prime factor in the local economy, is fiercely contested here. Because of this, many villages internally are far from stable, and more often, agricultural labourers, and scheduled castes lose their share of land. In village after village of the khadar region of Sonepat district, land is up for grabs. In Pawnera village, three powerful families grabbed 50 acres of panchayat land, and the rest of the panchayat land is now leased out for cultivation. Traditionally, this land was meant for grazing and other collective uses. In Atrena village, the surplus land of the panchayat had been distributed among the landless scheduled castes, but a few Rajput landowners have taken it by force. Panchayat lands in Murthal and Tokki villages have witnessed the same fate.

'Sixty Scheduled Caste families had been allotted at least four bighas of land each by the district administration, in the village, at the cost of Rs 1100 per bigha. We paid the entire money in instalments, and some of us took a bit of loan from the bank as well, to purchase tubewells. The powerful sardar landowners, who also have their land there, forcibly took control of our land three years ago, and since then we have been petitioning and protesting' complained Jagram, in Baghpur village of Faridabad district.

Girdawari is also the time when land changes hands illegally. It is the time, after every six to seven months, when government officials (like the patwari) measure the land, the crop, the output, and the ownership, in the khadar region. The rich and the affluent bribe the officials, and get the land of others in their records, which they then occupy. The legal remedy against this takes years. Villagers in Shekpur khadar, Mauja Bhud, and Jebabad Kheldi of Faridabad district, have had many bitter experiences in this regard.

People now do not live with khadar any more. They change it drastically, by raising artificial ridge-like embankments, by regulating and diverting channel flows, by constructing dams, by terminating the natural plant associations, and by reversing the agricultural and irrigation system. Somewhere a railway line crosses it, and sometimes the highway goes along with it. In places it is waterlogged, due to the channel flow and in others it is a low lying area with poor drainage. Hence, the khadar eco-system, which was always being perceived as the centre of the khadar environment and a system that had some boundaries for the environment, is no longer what it used to be. In its place is a dislocated region, full of disorder and discontent, and pauperization. Is khadar today turning into a *homacostatic* type system, which finally resists alteration in the environmental condition, takes its toll, and tries to exhibit the need to return to some equilibrium?

Hurly-Burly in the Coastal Seaboard

Fishermen blocking the sea at Kochi

Traditional fishing in Anjengo village

❦

Coastal Battles

Mechanized vs. Traditional Fishing

Fishery in Kerala passed through a difficult and violent phase in 1993. Burning trawlers, picketing and blocking roads and rail traffic, hunger strikes, mass rallies, police firing, etc. were all part of a struggle launched by traditional fisherfolk and their organizations. This signified the deepening crisis and emerging conflicts in the fishing sector. The demand for a ban on trawling during the monsoon made sense, given the solid evidence of the damage caused by over-fishing, the destruction of fish-breeding sites, and the depletion of marine stocks.

The fishermen's struggle had gained strength with the *Kerala Swatantra Matsya Thozhilali Federation*, the unions affiliated to the Centre of Indian Trade Unions, the All India Trade Union Congress, the Communist Party of India (Marxist–Leninist) and the Indian National Trade Union Congress forming a Joint Action Committee to address certain emerging issues. A spokesperson of the Joint Action Committee said, 'Now there is a virtual war between traditional fisherfolk on the one hand and mechanized trawlers, boat owners, private companies, export houses and upcoming multinationals on the other. This war will continue.'

Disquiet had spread all along the coast. Traditional fisherfolk from Anchuthengu, near Thiruvananthapuram burnt fourteen trawlers, seized four others and held hostage a boat operator. In retaliation, trawler owners took four fishermen as hostages. In June, trawlers were burnt or destroyed in Kollam and Kochi.

Felix, a traditional fisherman in Thumpoly village of Allappuzha district and now involved in ring-seine fishery, said, 'I have been fishing in the sea for twenty years. Now I can fish only for four to five

Published in *Frontline*, 24 September 1993.

months in a year because there is no fish for the remaining months. If I don't go continually for fishing, it is a severe loss for me. What will happen to my family and me? The church and some of the political parties provide us some relief. But we do not want relief. Now we have started burning trawlers. We have burnt eighteen trawlers so far. And that is the only way in which we can save our future.'

Heavy rains lashed Kochi on the morning of 16 July 1993 when thousands of traditional fisherfolk blocked the Thoppumpady—Ernakulam road for several hours, in response to the call given by the Joint Action Committee. At the same time, hundreds of traditional boats were used to blockade the sea. As the fishermen raised slogans, Thomas Kocherry, Chairperson, National Fishworkers' Forum, echoed their militant mood: 'We will intensify our struggle in the sea, as well as in the field. We will not allow mechanized trawling in the whole monsoon. We will, by all means, force the government to defend the interests of traditional fisherfolk.'

The issue that sparked off the fisherfolk movement was monsoon trawling. The issue has been surfacing year after year for almost two decades now. In 1993, the issue acquired a new dimension, as mechanized trawlers continued with their operation even after the state government formally announced a trawling ban from 15 June to 15 July. The traditional fishermen saw a collusion between the fishing lobby and the state government and feared they would lose whatever they had achieved in the course of their long struggle.

It was in 1979–80 that the traditional fisherfolk and their organization raised their key demand for a trawling ban during the monsoon. They also demanded effective enforcement of a trawler-free coastal fishing zone, marked exclusively for traditional fishermen operating non-mechanized craft, a total ban of purse-seine fishing in Kerala's coastal waters, and social welfare measures for fishermen. The state government enacted the Marine Fishing Regulation Act in 1980, which established a legal framework for zoning the coastal waters into areas reserved exclusively for traditional fisherfolk using non-mechanized craft, and areas further from shore for trawlers and purse-seiners. However, this law was not implemented. Only in 1989, after the *Matsya Thozhilali Federation*'s struggle, did the state government issue a notification under the Act, banning bottom trawling within the territorial waters (12 nautical miles). Simultaneously, a notification was also issued, prescribing conditions of minimum length, minimum horse power, safety applications, and so on, for boats permitted to fish in the monsoon beyond the territorial waters. From

1990, the state government announced a ban on monsoon trawling. Owners of mechanized trawlers and their association have been unsuccessfully challenging the ban and other restrictions every year.

In July 1992, the Kerala High Court issued an interim order staying the conditions imposed on fishing boats going beyond the territorial waters. The stay was initially for a week, but was extended from time to time. Thereafter, the Union government was also included in the petition to examine whether the central acts or rules, such as the Merchant Shipping Act, imposed any restrictions on boats going beyond territorial waters. On 25 March 1993, the high court finally struck down the state government's notification, largely on the basis that the state government had no jurisdiction to make laws beyond the territorial waters.

In this conflict, the Commerce Ministry of the Government of India took a stand in favour of mechanized trawlers. A ministry communication to the Kerala government said, 'There shall be no objection to harvesting of prawn resources during the months of June–August. A ban on monsoon trawling may, perhaps, not serve the purpose.'

A new situation thus emerged. Despite the ban on monsoon trawling in territorial waters, the mechanized trawlers were continually catching fish and prawn in the territorial waters only. Purse-seine fishing, a method of encircling whole shoals of pelagic fish, is also being resorted to by mechanized boats despite the ban.

For the government of Kerala it was not physically possible to police the 590 kilometre long coastline of the state during the monsoon. Neither was the Centre ready to monitor mechanized trawling or purse-seining beyond territorial waters. Rules under the Marine Shipping Act require compulsory registration and inspection of boats going out to sea. In particular, the Marine Shipping (Registration of Indian Fishing Boats) Rules, 1988, require compulsory registration of all mechanized boats. Similarly, section 435 (a) of the Act says that no Indian boat can ply in or proceed to sea without a valid and subsisting certificate of inspection. In particular, there are special requirements of life-saving equipment. Despite these rules being in force through a state government notification since 1989, allegedly not a single boat had a certificate of inspection, and very few boats had a certificate of registration.

For almost three months, the state government maintained silence on the high court verdict. However, pressed by the fisherfolk movement, the government announced in the state assembly on 1 July 1993 that it would appeal in the apex court against the verdict and take steps to enforce the ban on trawling. Neither announcement has been implemented till date. The fishermen's federation also filed a special leave petition in the apex court.

Meanwhile, to mollify the traditional fisherfolk, the state government appointed many committees on this issue: D. Babu Paul Committee in 1982, A.G. Kalawar Committee in 1985, Balakrishnan Nair Committee in 1989, and again in 1991. The Babu Paul Committee recommended on backwater fisheries that an area of two to three square miles be declared a fish sanctuary at important barmouths. No fishing, especially stake net and Chinese dip net fishing, was to be permitted within this notified zone. The stipulated cod ends of trawl and stake nets should not have mesh size smaller than 35 mm, and this was to be strictly enforced. No more new fixed engines would be licensed and all unauthorized stake nets had to be removed. For marine fishing the Committee recommended that the stipulated cod ends of trawl and stake nets should be enforced. No more shrimp trawling boats for inshore waters were to be allowed and all fishing craft had to be registered.

The Kalawar Committee suggested measures for the conservation and management of marine resources, including a reduction in the number of trawling boats and motorized and non-motorized craft, and the halving of the number of stake nets and Chinese dip nets (including unlicensed ones) in the backwater areas. Purse-seiners could not be dispensed with, according to the committee.

The Balakrishnan Nair Committee suggested, in the interest of conserving marine resources, a ban on trawling in the territorial waters of Kerala during June, July and August. No area was to be exempt from the ban. The impact of this measure on the conservation and optimum utilization of the resource was to be closely examined in the following three years (from 1989). In this period, a team of experts would study the impact of the ban on the status of resources, the socio-economic aspects of the fishermen's lives and the fishing industry.

The Nair Committee in 1991 noted with distress that since the ban on trawling was not imposed for the entire monsoon, it was not likely to produce any perceptible impact on catches. Although an increase had been noticed in the overall landings during 1988 and 1989, the committee had no conclusive evidence that this increase was a result of the restricted ban. The coverage of the ban over the years 1988, 1989, and 1990 was also not uniform over space and time. The ban period was reduced in 1989 and further reduced in 1990. The committee submitted that in the absence of adequate data, which should have been systematically collected after imposing the ban on trawling during the monsoon, it was difficult to arrive at definite conclusions regarding conservation of fishery resources along this coast as a result of the ban.

The state government officials took the stance that the problem of depleting fish catch could be resolved by allowing mechanized trawling only beyond the territorial waters, with a ban on night trawling, purse-seining, ring-seining, pelagic trawling and mid-water trawling, and a temporary ban on monsoon trawling except the Saktikulangara–Neendakara area, motorization of artisanal craft, and so on. But there were innumerable complaints of violations of these restrictions. According to *Kerala Fisheries, An Overview 1987*, 185 mechanized boats were impounded in 1984–5.

The All India Federation of Mechanized Fishing Boat Owners' Association disagrees with the view that monsoon trawling adversely affects the ecology. According to its spokesperson, there is no scientific proof that a ban on monsoon trawling increases marine production. On the contrary, it results in a major drop in the foreign exchange earnings from prawn export and increase unemployment. Besides, fishing in the coastal commons was a fundamental right of the mechanized boat-owners. The Association alleged that traditional fishermen attacked mechanized operations even beyond the territorial waters.

A comprehensive census of fishermen and fishing craft, carried out by the Kochi-based Central Marine Fisheries Research Institute (CMFRI) in 1980, found that Kerala had 99,820 fisher families with 6,40,000 members and 1,31,000 active fishermen. There are a large, steadily increasing number of boats, trawling boats, gill-netters and others. There is an unprecedented increase in the motorization of traditional craft as well.

According to the Balakrishnan Nair committee, the present estimates recorded a dismal picture. Several among the 4225 mechanized boats were unregistered. In addition, about 750 mechanized boats from neighbouring states operated in Kerala waters. Country craft, which include plane built boats, dugout canoes, catamarans and plywood canoes, numbered about 35,800, of which at least 10,850 were motorized. According to a recent estimate, there are about 2400 ring seines, besides more than 40,000 licensed and unlicensed fishing nets in the backwaters. The banned ring seine, which became very popular, was in effect a miniature purse seine, but considerably bigger than the traditional encircling net. It may be noted that the maximum number of ring seines recommended for Kerala's coast is 300 and that too, as a replacement of all existing traditional gear.

John Kurien of the Centre for Development Studies in Thiruvananthapuram reckoned that the harvest of nearly all important demersal (seabed dwelling) declined between 1971 and 1985, largely because of excessive or destructive fishing—particularly, the use of trawlers. Five major factors contributed to excessive fishing. First, the introduction of mechanized boats and percentage profit exporting. Rapid entry was facilitated by free access to the sea. Secondly, the use of the bell-shaped net for both demersal or bottom pelagic and surface dwelling fish. Thirdly, the introduction of trawlers into Kerala coincided with the rise in demand for prawns in the international market. From a commodity which once provided manure for coconut palms, prawns became the 'pink gold' of marine exports from India. Fourthly, to foster modernization, the state of Kerala actively promoted investment in the fishing sector, instituting many attractive subsidies. Lastly, the pressure exerted by an increasing number of fishermen within the limited area of the Kerala coastal waters. The active fishing population has been increasing by about 2 to 3 per cent per annum. In 1961, each fisherman had, on average, 16 hectares of coastal commons to fish. By 1985, the number of fishermen had increased by 65 per cent reducing the average coastal commons per fisherman to 9 hectares.

T.R. Thankappan Achari from the Fisheries Research Cell estimated that the share of the mechanized sector in fish production in Kerala increased from around 30 per cent during 1976–85 to 38 per cent in 1986–90, while the artisanal sector correspondingly dropped from 70 to 52 per cent. At the national level, fish production registered an annual growth rate of 3.2 per cent between the years 1971–5 and 1986–90. Kerala and Andhra Pradesh were the worst performers, with 1.3 per cent and 0.9 per cent, below the national average. On the export front, Kerala's share has been sliding.

On the consumer front, fish prices were rising in the state, as fish production and distribution did not keep pace with local demand. The administrative reports of various years from the Department of Fisheries note that in 1961–2, the beach price of prawns was only Rs 240 a tonne. In 1971–2, 1976–7, and 1984–5, the prices went up to Rs 1810, Rs 7260 and Rs 14,120 a tonne respectively. The price escalation was reflected even for the domestically consumed species of oil sardines and mackerels. The prices of the former in 1961–2, 1971–2, 1976–7, and 1984–5 were Rs 90, Rs 380, Rs 850 and Rs 1080 a tonne respectively. The price of mackerel, considered the poor man's source of protein, went up from Rs 340 a tonne in 1961–2 to Rs 2790 in 1984–5.

Fishermen are subject to new hardships. K.V. Raphael from Puthuvypin village of Ernakulum district, who worked on a traditional long boat, recounts how competition has compelled him to use outboard engines in boats for fishing. The engine, boat and the net cost about Rs 5 lakh. The value of 70 to 75 per cent of the catch goes as loan repayment, fuel cost and to middlemen. Raphael is still in debt of more than Rs 20,000. Ambika, from Pallithodu village in Allappuzha district, sketched the plight of fish-dependent families now in distress, 'My father is a fisherman. I have three sisters and one brother. Five years ago, our whole family was involved in drying fish. Then came a time when there was no fish to dry. Now whatever fish my father catches, he immediately sells in the market. There is no surplus fish to store or to dry. So, we remain idle. The same is the case with my sisters and brother, and most of the village women.'

The dismal situation of fishery in Kerala has prompted the formation of a Fish Resources Protection Council, a citizen's body, under the chairmanship of former Chief Justice of the Supreme Court, V.R. Krishna Iyer. The council is organizing a statewide campaign on environmental and consumer issues in this sector. Now traditional fisherfolk, their organizations and most of the fishery experts are unanimous in demanding a ban on monsoon trawling and other restrictions on mechanized fishing in the coastal waters. The Fishermen's Federation asserted that as a result of the trawling ban, fish landing has improved, especially in the traditional sector. The CMFRI figures showed that in 1989 and 1990, Kerala marine fish landing crossed six lakh tonnes. The highest figure was 4.5 lakh tonnes in 1973.

Traditional fishermen are also demanding a new legislation, which will give the right to own and operate fishing implements exclusively to actual fishermen, enforce orders issued under the Kerala Marine Fishing Regulation Act, and allow them to collect arrears from exporters and boat owners towards welfare funds and so on. But the central government, the state government, and the mechanized fishing sector are going ahead with their plans of over-exploitation and massive exports.

If Only the Sea could Tell

Anjengo is a large village of traditional fisherfolk, 46 kilometres away from Thiruvananthapuram, surrounded by Mampally, Poothura, and Thazhampally, also villages of traditional fisherfolk. Violent incidents occurred in this otherwise tranquil region in November 1993, which indicated that the conflict between traditional fisherfolk and the owners of mechanized boats or trawlers over fishing rights in the sea has reached a new and difficult phase.

On 28 November, the fisherfolk, church leaders, and Kerala Swathanthra Matsya Thozhilali Federation's activists held meetings in both Anjengo and Mampally villages. The villagers were tense and worried. Father Wilfred in Mampally village said, 'We are discussing the possibility of forming a village-level association of fisherfolk, to defend ourselves.' In the night, when many trawlers approached the village to attack the fishing boats and fisherfolk and to release two trawlers which had been 'captured' in the territorial waters by the villagers, dozens of fisherfolk sailed out in their small boats to guard the captive trawlers. At this, the approaching trawlers changed course.

For more than 20,000 families of Mukkava caste in Anjengo, Mampally, and other surrounding villages, the only occupation is fishing. In this area, popularly known as the 'fish bank', some important and exportable fish species, cattlefish and prawns are found. Mechanized boats or trawlers from Shaktikullangara, Kavanda, Neendagara, Aravilla, Kuruvipara, all in Kollam district, and their operations in the sea, are adversely affecting the share of these villages in the coastal common. The trawlers not only violate the monsoon trawling and night trawling bans in the territorial waters; they now want to establish more control over the area.

Verghese, a traditional fisherman in Anjengo village, narrated how

Published in *Economic and Political Weekly*, 26 February 1994.

the trawlers and their men attacked and destroyed their boats and nets in the sea. Sometimes they captured their boats and nets. Nowadays they go fishing fearfully. Whenever they spot a trawler, they rush back to the shore. Paniadima said: 'Out of twelve months, only five to six months are the fishing season for traditional fisherfolk. October, November, and December are the real working months for us. But in these months the sea is not ours and has been put beyond our reach. Nowhere are there marine forces in this area to protect us.' J. Thobai's outboard engine, net and other instruments, purchased under a bank loan, all got destroyed in the recent attacks.

The conflict between traditional fisherfolk and mechanized boats regarding rights over the sea, having emerged over many years in the maritime states, has degenerated rapidly, with episodes of violence from different parts of Kerala and Tamil Nadu. On 10 August 1993, violent clashes occurred in Inayam Puthenthurauil in Tamil Nadu, where many traditional fishing crafts and trawlers were burnt. Similar conflicts took place in September 1993 in Muttam, Tamil Nadu. On 10 October 1993, traditional catamaran fishworkers in Kerala captured twenty trawlers from the territorial waters.

In early November, Eastus, a fisherman, recalls, 'It was raining. We were fishing in our small boats with gillnets. We did not hear the sound of the trawlers which suddenly surrounded us and people attacked us with stones, knives, and sticks. They captured our boats and destroyed our nets. Paniadima, Stephen, and Vincent jumped into the sea to save their lives. But Antony and two other fisherfolk were taken to Shaktikullangara and kept hostage by the trawler-owners. They were badly beaten up and had to be hospitalized.' On 8 November, at the intervention of the local authorities, police and the church, Father G. Stephen in Anjengo and Sister Sherly in Mampally village went to Shaktikullangara to mediate and seek the release of the hostages. According to Sister Sherly, 'As soon as we reached the place in two taxis, various people, armed with rods, attacked us. Our taxis were set on fire, our belongings snatched, and all of us were locked in a room. The local police posted there were of no help. Some of us were seriously injured.' The group comprising fisherfolk, social activists and the church people were released late in the night, but the captured boats of the fisherfolk were not given back. They were burnt before their eyes.

After some days, the trawlers re-entered the territorial waters in the same area, defying the ban on night trawling. This time, the fisherfolk caught hold of two trawlers along with nine workers. The workers were freed within some hours on humanitarian grounds, but the fisherfolk refused to release the trawlers without any concrete assurances on the part of administration officials and trawler associations, but the administration was nowhere to be seen. Some time earlier, at a meeting with the Kollam district administration, it had been decided that five marine force boats with modern equipment would be posted in the area, but not even one was posted in the problem area. Later on, the district administration expressed helplessness and merely urged the fisherfolk to be vigilant and catch violating trawlers. This worsened the situation.

Traditional fisherfolk and the Kerala Swathantra Matsya Thozhilali Federation perceive this new situation in the context of the new economic policy. The president of the federation said that when the government aggressively pursued the deep-sea fishing policy and started many joint ventures to exploit the fish resources in the coastal areas, the battle in the sea was bound to aggravate. The sea told the stories of these battles. After the November incident, the Latin Catholic Church initiated village-level organizations of traditional fisherfolk in some areas. Father Selvadas said of this initiative, 'If the government is not enforcing the laws, we will make our laws and enforce them on our strength.'

Fishing Harbour
Caught in a Time Warp

Thankassery harbour in Kollam district, which was being developed ostensibly for the benefit of traditional fisherfolk, witnessed police firing on 18 June 1993 after the anchoring of mechanized boats near the breakwater at Thankassery led to a clash. While mechanized boat-owners and trawlers considered it their right to have their boats anchored here, the traditional fisherfolk saw it as yet another assault on their already shrinking space in the marine sector and tried to resist it.

Thankassery, Badi, Mottakara, Portkollam, and Parlitotam are fishing villages near the harbour. On the evening of 17 July a large number of villagers met in Kollam for over five hours to analyse the impact of the harbour on artisanal fisherfolk. Most of the fisherfolk expressed alarm over the hijacking of the harbour by mechanized trawlers. But some of the artisanal fisherfolk, mostly from Thankassery village, which is adjacent to the harbour, wanted the construction of the harbour as well as the anchoring of mechanized boats and trawlers to continue. Winston of Thankassery explained that mini-trawl units had emerged in a big way in this area in the past two to three years. Now some of the traditional fisherfolk also operate the mechanized trawlers. They expected the harbour to generate employment and other commercial activities in the village.

The villagers' meeting, however, finally decided to oppose the operation of mechanized boats and trawlers from Thankassery harbour. A.J. Vijayan, editor of *Waves* (a fortnightly from Thiruvananthapuram on fisherfolk), who also addressed the meeting, said, 'It has been decided that only artisanal fisherfolk should be allowed to operate from Thankassery harbour. The district administration should

Published in *Economic and Political Weekly*, 18 September 1993.

ensure necessary regulations regarding this. If this is not done, a serious law-and-order problem will emerge in the whole area.'

The technological changes during the last one decade have drastically changed the artisanal marine fishing fleet in Kollam district. A result of these changes are frequent conflicts among traditional fisherfolk as well as against trawlers and mechanized boat-owners. The northern part of the district, from Neendakara to Azheekal, abounds in 60–70 feet 'thanguyallom' and 20–25 feet 'dinghies'. The fishing harbour at Neendkara is the centre of thousands of mechanized trawlers during the monsoon months. A large number of migrant fisherfolk from Kanyakumari have also settled here permanently and usually use plywood boats and hook-and-line fishing in the deeper waters. South of Neendakara, Kollam town fishery, from Thankassery to Pallithottam, is dominated by hundreds of motorized plywood boats and fisherfolk operate a number of small gillnets. A group of migrant fisherfolk at Jonapuram use hook and line. South of Kollam town is also marked by non-motorized near-shore fishing operations. But overall, the mechanized sector has a devastating impact on traditional fisherfolk in the district. The operation of trawlers at Neenadakara, as well as the ring-seine units further north, severely affect the traditional sector. Mini-trawl units have emerged in a big way in northern Kollam. Many residents of these fishing villages are now forced to work in the mechanized sector.

A survey by the South Indian Federation of Fishermen Societies in 1991 found that a striking feature of the 'thanguvallom' belt of Kollam was the total absence of beach landing. Sea erosion is a severe problem. This forced the fishermen of Azheekal, for example, to travel nearly 25 kilometres through the canal to reach Neendakara. The fuel expenses for just reaching the sea and returning home after fish sales at Neendakara, are Rs 250 per trip. The underwater exhaust of kerosene outboard motors, fitted on the 100-odd 'thanguvalloms' that undertake the trip each day, pollute the backwaters, making the fish catch from the backwaters smell of kerosene. After the marketing activities shifted to Neendakara in 1987, a number of small vendors who used to purchase fish on the beaches have been deprived of their livelihood.

A project costing Rs 1859 lakh was prepared for the integrated development of Thankassery fishing harbour under the Eighth Plan, 1990–5, supposedly intended for the traditional sector. Other components of the project, apart from the fishing harbour, are motorization of 400 existing craft with outboard motors, introduction of 400 marine plywood canoes and twenty dory fishing units. The project, scheduled to be completed in five years, was expected to generate an additional annual fish production of 20,467 tonnes. With the harbour complex, the actual fishing days per year would increase to 250. If the breakwater were extended to Pallithottam, port officials pointed out, apart from the ten fishing villages of Thankassery, fisherfolk around Pallithottam could also berth their vessels. It was also expected to increase the number of craft from 1200 to 1600, with an increase in gear and outboard engines.

The establishment of harbours for fishing needs was, in fact, an important programme of the modernized export-oriented model of development of the marine sector in Kerala. As part of the programme, the construction of Vizhinjan as a harbour for deep-drafted vessels started in 1962. Harbour projects at Neendakara, Azheekal and Mopla Bay, to cater to mechanized boats, were also envisaged. Fishing harbours at Cochin were commissioned in 1978 under the central sector. Neendakara was commissioned in 1988 as a harbour catering mainly to mechanized boats, but with provisions for deep-sea vessels. Mopla Bay was taken up in 1963 as an Indo–Norwegian project. The Azheekal project also was partially completed. Of these ten harbours, nine were planned to cater to mechanized boats or deep-sea vessels.

At Muthalapozhi, Kayamkulam, Thottappally, Chettuvai, and Kasaragod the possibility of establishing fishing harbours was being explored, but meanwhile, the utility of the development of the harbours for fish catch was increasingly proving negative. The Balakrishnan Nair Committee in 1989 concluded that penaeid prawn landings in Kerala, particularly and at the Neendakara and Cochin Fisheries Harbour centres during 1975–88 showed considerable fluctuations over the years. At Neendakara, with about 57,000 tonnes in 1975, the maximum during this period, the landings registered a downward trend with periodic jumps at lower levels. In 1988, the catch was a mere 9100 tonnes in spite of the increasing trend in the landings at Cochin Fisheries Harbour and in the whole of Kerala in the late 1980s. The decline was due to fishing pressure, together with indiscriminate fishing of young ones. John Kurien and T.R. Thankappan Achari noted: 'In the main prawn landing centre in Kerala, in Neendakara the catch per unit declined from 83 kg/hr of fishing effort in 1973 to 20 kg/hr in 1984.'

Traditional fisherfolk in Thankassery harbour area were alarmed over the proposed construction. The villagers' meeting and police firing were but a foretaste of a major conflict. Elias, who stayed on the banks of Asthamudi lake in Kollam district, opined, 'The harbours are built in natural and safe areas. These areas have been traditionally used by the traditional fisherfolk. But now these are being converted into harbours. After the harbour construction, we will be displaced. We are caught in a time warp, not knowing which way we are heading. At least we should be ready to ask who will actually benefit from these so-called developmental projects.'

Enter, the Big Fish
New Deep-Sea Fishing Policy Draws Flak

In the current liberalized and globalized environment, domestic and foreign corporate investments and joint ventures, initiated on a massive scale for deep-sea fishing in India's exclusive economic zone (EEZ), are threatening the indigenous fishing sector. Representatives of associations of traditional fisherfolk, small-scale mechanized fishing boat operators, fish traders, processors and exporters met in Kochi in June 1994 and formed the National Fisheries Action Committee Against Joint Ventures. The meeting demanded that the government cancel all licences issued to joint ventures and stop issuing licences for deep-sea fishing. It also decided to observe 20 July as a Black Day. The meeting called an indefinite all-India fisheries strike from 23 November, and a blockade of foreign fishing vessels visiting the ports for refuelling.

The struggle against the new policy, which also involved opening up the seas to foreign operators, multinationals and joint ventures, had gained momentum in the months preceding the meeting. On 4 February, the National Fishworkers' Forum (NFF), in collaboration with the Small Mechanized Boat Owners Association, the Association of Wholesale Fish Merchants and thirty-one other organizations, staged an all-India fisheries bandh, and not one fishing boat, artisanal or mechanized, went to sea in any of the coastal states. The major wholesale and retail fish markets remained shut.

However, the government had obviously decided to ignore the protests. According to S.L. Kapur, Secretary, and C.K. Basu, Joint Secretary, Ministry of Food Processing Industries, India's vast marine and inland water resources had been traditionally tapped by local fishermen to cater for domestic demand. Over the past decade or so,

Published in *Frontline*, 26 August 1994.

the organized corporate sector had become involved in the preservation and export of coastal fish. Despite this, India's fishery resources remained grossly under-utilized. Therefore, in the new environment, domestic and foreign corporate investments were being encouraged, both for inland farming and for marine fishing. Large Indian and multinational corporations had set up shrimp culture, deep-sea fishing and seafood processing projects. Most of these projects, according to Basu, were being set up with the export market in view.

The officials also noted that many leading industrial houses now had a presence in the deep-sea sector. These included Hindustan Lever, Dunlop, ITC, United Breweries, the Thapar group and Amalgam Foods. Important joint venture collaborations included Chevanne Merceron Ballery and Cofrepeche, France; Consolidated Sea Food Corporation, USA; Toyo Kosuisan, Japan; and Marine Corporation, South Korea. Though the Union Ministry of Food Processing Industries listed only forty joint ventures in deep-sea fishing, the action committee alleged that the government had issued 129 licences for joint ventures, which included huge foreign companies like Mitsubishi of Japan, Kalis of Australia and AM Produkte of Germany.

With a coastline of 7500 kilometres and an EEZ of 2.02 million square kilometres, India has an estimated marine catch potential of four million tonnes. Of this, half is reportedly within the coastal area, that is, up to a depth of 50 metres. This is where the traditional fisherfolk ply their craft. But trawlers, mechanized boats and their destructive fishing technology increasingly corner the catch. The marine catch and exports of the mechanized sector have increased steadily, both in terms of quantity and value. The catch, including prawn, shrimp, tuna, cuttlefish, squid, octopus, red snapper, ribbon fish, mackerel, lobster and cat fish, was 1.09 million tonnes in 1970–1 and 2.57 million tonnes in 1992–3. Exports have risen from 99,700 tonnes in 1989–90 to 2.24 lakh tonnes in 1993–94. The compounded growth rate during this period was 22.4 per cent. In value terms, earnings shot up from Rs 598 crore in 1989–90 to Rs 2,320 crore in 1993–94. But the share of the traditional fisherfolk in this has declined. Domestic per capita consumption was also static—a mere 3.5 kilograms a year, as against the world average of 12 kilograms.

Region-wise, the EEZ off the west coast forms about 27.8 per cent of the total and that around the Andaman and Nicobar Islands about 29.6 per cent. The government aims to exploit this deep zone as fast as possible to increase exports and foreign exchange earnings. A working group constituted by the government on 'Revalidation of the potential of marine fishery resources of EEZ of India' suggested, in February 1991, that the revalidated potential yield of 3.9 million tonnes in the Indian EEZ consists of three major components—the inshore fishery resources of up to 50 metres depth all along the east and west coast contributing 2.21 million tonnes; offshore and deep sea resources from 50–500 metres depth contributing 1.395 million tonnes; and oceanic resources, consisting of tuna and allied fishes and shark, 0.295 million tonnes. 'The unexploited and under exploited resources beyond 50 m depth are the main target for the future developmental strategy... . Even for exploitation at 50 per cent level of additional resources beyond 50 metre depth, a total of 2,630 vessels of different sizes and types are necessary.' According to the report, deep-sea vessels are 23 to 40 metres in size and capable of harvesting 120 to 2,000 tonnes a unit, totalling 0.83 million tonnes—half the potential of the EEZ.

A new phase thus started with a series of incentives and schemes offered by the government to encourage joint ventures, foreign vessels and multinationals in the Indian deep seas. Under the liberalized norms (General guidelines on joint ventures; Broad guidelines for joint ventures in deep-sea fishing; Broad guidelines for leasing of foreign deep-sea fishing vessels; Broad guidelines for test-fishing in the Indian EEZ; Benefits to 100 per cent export-oriented units, and so on), up to 51 per cent foreign equity participation was permitted for joint ventures.

Companies were allowed to do test-fishing with foreign collaborators to facilitate pre-investment studies. Both new and second-hand foreign fishing vessels could be leased. Foreign crews were permitted for joint ventures. As a special incentive, the transfer of catch on the high seas was permitted for deep-sea fishing vessels. Financing was available from the Shipping Credit and Investment Corporation of India (SCICI) for the acquisition of vessels and gear.

The government had earlier tried various options to develop deep-sea fishing, with disastrous consequences. In the early 1970s, shrimp trawlers were imported from Mexico and the US and then transferred to fishing companies in the public and private sectors. In 1977, the government introduced the system of licensing for acquisition of vessels by Indian fishing companies. It also permitted chartering of foreign vessels by Indian companies. Again, the government introduced licensing of 100 per cent export-oriented units. Under this scheme, Indian companies were exempted from the *pari passu* clause of placing orders for Indian-built vessels. According to an official, till 1990 the government had licensed the acquisition of 750 vessels—312 vessels under the general scheme, 129 under charter obligation and 309 under an export-oriented unit scheme.

A note of the Ministry of Food Processing Industries indicates that 180 deep-sea fishing vessels were in operation in 1994. However, the policy of chartering of foreign fishing vessels had been largely phased out, and there was greater emphasis on joint ventures.

The FAO Mission for the Development/Diversification of the Deep-Sea Fleet at Visakhapatnam in its summary report of 1992 had exposed the hollowness of previous government policies. On deep-sea fishing in Visakhapatnam, the report said, 'The DSF fleet has presently 180 trawlers ... and a total power of 81,000 horse power. In spite of this, its yearly shrimp production has always been small and in recent times, has stagnated around 1000–1500 tonnes ... . This fleet is owned by 96 enterprises. Its low productivity gives the impression that probably no more than about 10 per cent of them are financially sound today ... . An over-capacity for production, which has induced a situation of strong over-fishing of the target resources ... . In conclusion, one could say that the main problem of the DSF is not so much its capital and operation costs, which have been generally fair by developing country standards. The primary problem is, by far, the situation of over-investment in the shrimp business and subsequently, of over-fishing of its target resources. Therefore, the priority need for this fishery is not further development but resource management.'

Thousands of tonnes of by-catch from shrimp trawls are being discarded annually on the north-eastern coast. The Bay of Bengal Project had made some observations regarding a few trawlers operating off West Bengal. The ratio of shrimp to by-catch was 1:14.7 for the large trawlers and 1:23 for the small trawlers. This by-catch had a high proportion of immature fish, whose capture affects the next day's catch. And a majority of the discarded species are those that form part of the catch of small-scale fisherfolk.

T.R. Thankappan Achari of the Fisheries Research Cell, Thiruvananthapuram, and A.J. Vijayan of the Kerala Swathanthra Matsya Thozhilali Federation, have elaborated on the likely consequences of foreign investment in the EEZ of India. According to their perception, after exhausting the valuable species, which constitutes scarcely 15 per cent of the total deep-sea potential of India, these new entrants would switch to the cheaper varieties even for making products like fish meal, for which there is a good market in the west, and to feed pets, poultry, pigs, and so on.

The fishing technologies of foreign vessels are more suited to temperate waters, where large stocks of each species are available. When similar technology is applied to tropical seas, where comparatively smaller stocks of many different species of varying sizes are found, it definitely leads to faster depletion of stock. Fish is a renewable natural resource, but the larger varieties like tuna, perches, shark, and catfish, once depleted, take many years to recover. Each fish variety in a particular region is a link in the chain of total fish resources. The destruction of one link could break and collapse the chain. Besides, the promotion of deep-sea fishing will not substantially accelerate employment opportunities.

Just as mechanized boats ignore state government regulations on fishing in territorial waters and attack traditional fisherfolk, the use of foreign deep-sea vessels also leads to conflicts and clashes in many states. P.V. Kokhari, secretary, Shree Porbandar Machhimar Boat Association, complained that foreign vessels were supposed to operate 24 nautical miles away, but many of them often came very near the inshore area and violated the rules. They often damaged the nets and gear of the small fishermen. Often this was reported to the Coast Guard, but before their arrival these vessels pulled out.

Some deep-sea foreign vessels also catch and ship precious species of fish, shrimp and lobsters to foreign countries, unhindered throughout the year. Even the prohibition on rainy-season catch, which is a must for the regeneration of marine resources, is disregarded. It is therefore genuinely feared that under the joint ventures, Indian authorities will never be able to get a clear picture of the catch and catch values.

Significantly, some state governments have questioned the new policy. According to M.T. Padma, Kerala's Minister for Fisheries and Rural Development, the state government did welcome deep-sea fishing, but it could not go for deep-sea vessels because of financial problems. So they sought joint ventures. On the other hand, the Gujarat and West Bengal governments have strongly protested against this policy. Shashikant Lakhani, Minister of Industries, Cottage Industries, Mines, Gujarat, in a confidential letter addressed to Union Minister of State for Food Processing Industries, Tarun Gogoi, raised several questions: 'During the last few years the Ministry of Food Processing Industries has been licensing foreign vessels under the charter and joint venture programme to fish off the Indian coasts.

A large number of these vessels are in fact fishing off the Gujarat coast during the last few years. The state government is not consulted or even informed ... . The details of the fish catch and foreign exchange realised, etc., are not made known to the state governments and as such we are not in a position to evaluate the usefulness or desirability of such joint ventures in fishing. However, the fishing activity by foreign vessels has caused great concern and fear among the traditional Gujarat fishermen. Apart from the damage done to the nets and fishing gears of the Gujarat fishermen, there is acute anxiety regarding the destructive methods employed by the huge vessels ... . I, therefore, request you to stop the foreign vessels from fishing in the Indian waters under the charter and joint venture programme immediately.' The West Bengal Fisheries Minister, Kironmoy Nanda, recently announced that the state would not allow multinationals and joint ventures in the fishery sector.

Amidst all this, the NFF, a united body of fish workers' trade unions and a major constituent of the Action Committee, also sought a new policy approach to deep-sea fishing. As Thomas Kocherry, chairperson, NFF, clarified, fish workers were not against a policy of deep-sea fishing aimed at the sustainable development of the sector, sharing benefits with fishermen and the national economy. The policy, according to him, had to ensure the expansion of the ambit of operations of the small fishermen to deeper waters. Enterprising fishermen had to be encouraged and supported to move into offshore waters. The policy had to also ensure liberalized central subsidies and credit for small fishermen who ventured into the deep seas. It had to lead to increased supply of fish for domestic consumption. The government had to confer legal rights and reserve exclusive fishing zones for small-scale artisanal fishermen, at least up to the contiguous zone. Annual fishery management plans with estimates of total allowable catch, introduction of quota systems, fishing holidays and surveillance need to form part of resource management.

Artisanal fishermen have proved that deep-sea fishing is feasible with very low investment and technology, based solely on traditional navigational and catching skills. The artisanal fishermen of Thoothoor village in Kanyakumari district of Tamil Nadu manage almost the entire shark landings on the west coast. With around 350 boats, their catch averages 750 kilograms a week. The annual sum is 13,000 tonnes or more, almost the same as that of an even more dispersed fleet from five or six countries operating in the Bay of Bengal. The South Indian Federation of Fishermen Societies (SIFFS), an apex body of three district-level fishermen federations of Kanyakumari, Thiruvananthapuram and Kollam, is trying to tackle the problems of these fishermen.

Thoothoor is less than 10 kilometres from the Kerala border. The old village is now divided into three—Thoothoor, Chinnathurai and Eraviputhenthurai—but the cluster continues to be called Thoothoor. There are at least three thousand fishermen and the crew size in each boat ranges from six to ten. The season starts after the south-west monsoon, by mid-August, with boats moving from Kanyakumari district to Mangalore, with brief halts along the Kerala coast. After registering with the Customs in Mangalore, they move north, fishing around different ports. By end-September most of the boats reach their final destination for the season and stay put till March or April. Half the boats reach Maharashtra, with the rest staying in Gujarat and Karnataka. Kurli in Maharashtra is a popular base, with over a hundred boats stationed there during the season. The fishing method is exclusively bottom long lining. The number of hooks ranges from 150 to 300, depending on baitfish availability and other factors. The baitfish is caught with hand lines and troll lines on the way to the fishing grounds. The entire continental shelf area is covered. Quite a lot of the fishing takes place on the edge of the shelf up to a depth of 300 metres.

According to V. Vivekanandan, chief executive, SIFFS, the shark fishermen face many problems as migrants to other states. While crores of rupees are spent in promoting deep-sea fishing, using large high-tech vessels, this is a group that has proved that deep-sea fishing is feasible on considerably lower outlays. Given the right kind of support, groups like the Thoothoor fishermen could commercially exploit offshore and deep-sea fisheries most cost-effectively.

The marine fisherfolk population of the country has grown at a faster rate than the overall population. For those dependent on inshore fisheries, the use of offshore/deep-sea resources is an important means of additional employment. But the government does not seem concerned.

The inshore resources are being exhausted by the advent of mechanized boats using methods like trawling and purse-seining; it is now the turn of deep-sea/offshore resources. It is ironic that on 20 July 1994, even as the National Fisheries Action Committee Against Joint Ventures (NFACAJV) was observing the Black Day to protest the promotion of joint ventures in India, Tarun Gogoi, the Union Minister of State for Food Processing Industries, announced in Delhi that there was nothing wrong with the deep-sea policy.

POSTSCRIPT

The one-day strike in February was followed by a two-day strike on 23 and 24 November 1994, more widespread and militant in its form and content. In some areas, small-scale fishworkers surrounded foreign vessels and prevented them from going to the sea; in others, people showed their support with rallies, sit-ins and fasts. Several members of parliament wrote to the Prime Minister in support of NFACAJV's demands, while some maritime state governments came out openly against the central policy. All the major central trade unions endorsed the strike, as did the popular social and environmental movements, of the country.

However, government's reactions were contradictory. A day before the strike, the government announced that it would implement a totally impractical and unrealistic 'corridor at sea' to reduce the conflict. The Minister for Food Processing Industries announced in parliament that he would issue a freeze on new joint venture licences, but even after this promise, the granting of licences continued. The minister appointed a review committee in February 1995. The composition of the committee, headed by P. Murari, former Secretary of the Ministry, and its terms of reference, lacked credibility. Hence, the NFACAJV boycotted its sittings.

As a result, when Thomas Kocherry, NFACAJV's convenor, began his fast in May 1995 in front of the birthplace of Mahatma Gandhi in Gujarat, it generated wide support across the country. In parliament, members belonging to political parties of all shades were unequivocal in their condemnation of government policies and lent their support to the struggle.

With the eight-day hunger strike, the movement gained new energy, and a combination of action programmes guaranteed public attention, public support and partial government action. Above all, it created an outlet for a change in the composition and terms of reference of the review committee on deep-sea fishing policy. This opened a new field in the fish workers' struggle.

The Murari Committee Report: A Milestone

The reconstituted Murari Committee, with 41 members, included 16 members of parliament, 6 representatives of fishing interests, all the fisheries secretaries from the coastal states and scientists and government officials, after visiting the coastal areas, in a sense virtually rediscovered the deep-sea fishing of the country. The Committee's report is the most far-reaching document for the management of deep-sea fishing prepared by any committee in the country since independence. The most important among its twenty-two recommendations were:

- All permits issued for fishing by joint ventures/chartered/lease/test should immediately be cancelled subject to legal processes as may be required.
- No renewal/extension of new licences/permits be issued in future for fishing by joint ventures/charter/lease/test-fishing vessels.
- All licences/permits for fishing may be made public documents and copy thereof made available for inspection in the office of the registered authority.
- The areas already being exploited by fishermen operating traditional craft or mechanized vessels below 20 metres size or areas which may be exploited in the medium-term future, should not be permitted for exploitation by vessels above 20 metres length except for Indian-owned vessels currently in operation, which may be given three years time to move out.
- Deep-sea fishing regulations should be enacted by Parliament after consulting the fishing community.
- Government should take decisions on the recommendations of the committee within six months.

On 23 February 1996, the government announced some interim measures based on these recommendations. The then food processing secretary declared that deep-sea fishing would be banned during the breeding seasons on the basis of recommendations of the state governments in their coastal areas, provided the states also imposed a similar prohibition in the territorial water trawling. However, he stated that the final decisions were pending 'which were being examined by several ministries'. But these decisions were neither taken nor the licences cancelled. This called for a new phase of struggle of fishworkers.

August 1996: A New Phase

On 7 August, Mumbai was busy, but with a difference. Amidst the traffic on the streets of Colaba and crowds in the markets and stalls, hundreds of fish workers, social and political activists, intellectuals and writers had banded together. Their leader, Thomas Kocherry, Chairperson of NFACAJV began an indefinite fast. They were demanding the cancellation of all licences issued to foreign industrial deep-sea fishing vessels to operate in the Exclusive Economic Zone and the implementation of the recommendations of the Murari Committee.

Thomas Kocherry commented in an interview that their intention was not to renew the agitation, but demand justice on the merit of the issue. Eighty lakh fisherfolk had been agitating against foreign fishing vessels for the past three years. But governments which worked for Enron or Cogentrix and responded to them in a matter of days, Kocherry reasoned, did not find time for the fishworkers and thereby naturally failed to meet the deadline of even six months, given by their own committee.

'If they harm the deep sea and fish like this, we will die. How can they even think of continuing with the corrupt practices of the past?', said one representative of a fishworkers' organization in Gujarat, while sitting on a dharna at Janpath, New Delhi. They were also joined by many prominent citizens, trade unions, peoples' organizations and voluntary agencies. This was probably the first time when all the prominent political parties and trade unions openly came out in support of the demand for cancellation of licences.

The entire coastline simmered with discontent in the following days. Everyday hundreds of fisherfolk fasted, and thousands marched out into the streets of Kerala, Karnataka and Maharashtra. In West Bengal, fisherfolk stormed the Coast Guard office at Haldia and staged an indefinite dharna. In Orissa, they held a rally and picketed the Chathrapur District Magistrate's office. Visakhapatnam witnessed a harbour blockade. All this and much more forced the Union Minister for Food Processing Industries to come down to the fasting shed at Machimar Nagar, Colaba, on the evening of 13 August.

14 August was a day of reckoning for the agitating fishworkers and their fasting leader. The minister gave a written statement which read, 'No fresh licence for charter, lease, testing and joint venture vessels will be entertained for Deep Sea Fishing vessels in our Exclusive Economic Zone ... Licences already issued will also not be renewed any further ... The Ministry will take steps towards implementation of the recommendations within a month including the issue of cancellation of licences after examining and resolving the legal and financial implications. The Ministry also proposes to convene an early meeting of experts in the field, the State Government

representatives and representatives of fishermen to discuss and evolve a comprehensive fisheries policy that ensures sustainable fisheries, upgradation of the skills of fish workers and the welfare of traditional fishing community.'

The agitation was called off. Hundreds of fishworkers marched on the streets of Mumbai, singing and dancing, swirling and pulsating.

Aquaculture
In the Midst of a Blue Revolution

The coastal belt of Tamil Nadu, Andhra Pradesh and Pondicherry has been experiencing signs of social and environmental upheaval as a result of the development of large-scale shrimp aquaculture projects, the so-called blue revolution.

The prawn farming boom has led to intense conflicts. On the one hand farmers and the fisherfolk have brought work in various upcoming prawn firms to a grinding halt. On the other hand, companies, with support from local rich people, police and administration, sought to commence production. The Union and state governments, however, were in a dilemma. Though committed to the development of shrimp aquaculture projects to cater for the export market, they now had to consider its ecological and social aspects as well.

The then Union Minister for Agriculture, Balram Jhakhar proudly stated in July 1995 that fish production touched an all-time high of 46.8 lakh tonnes in 1993–4. The export of marine products increased from 2.09 lakh tonnes in 1992–3 to 2.44 lakh tonnes in 1993–4. Foreign exchange earnings from these exports also increased substantially from Rs 1768 crore in 1992–3 to Rs 2504 crore in 1993–4. He accepted that if adequate steps were not taken to make commercial shrimp farming and aqua farming activities environmentally safe and eco-friendly, this rapid development could adversely affect the environment. Large-scale shrimp aquaculture practices could lead to multi-user conflicts, and to the salinization of productive agricultural land and fresh-water aquifers. Other side-effects of aquaculture included denial of access to basic amenities to coastal fishermen and fish landing centres.

The Ministry of Agriculture constituted an expert committee

Published in *Economic and Political Weekly*, 3 December 1994.

to suggest guidelines to make aquaculture projects more 'environment-friendly'. Environmental clearance was made mandatory for aquaculture projects. The committee's guidelines were to be given to the states and other end-users for agriculture. The ministry also asked states to survey and identify land suitable for aquaculture, specifically for shrimp, and to ensure that productive agricultural land was not diverted for brackish-water aquaculture.

The Marine Products Export Development Authority's composition-wise break-up showed that shrimp continues to be the major marine item exported. Of the thirty-odd items of marine products exported from the country, lobster, cuttle fish, squid, frozen fish and dried items like dried fish, shark fin and fishmaw form the other major items. Frozen shrimp export formed over one-third of the volume and two-thirds of the value in 1992–3. The export of 74,393 tonnes yielded Rs 118.26 crore. The major markets were Japan, USA, UK, Italy, Spain, Netherlands, and France.

India has long had a traditional rice/shrimp rotating aquaculture system, so that rice would be grown part of the year and shrimp and other fish species cultured for the rest of the year without using processed feeds, chemicals and antibiotics. However, by the early 1980s, with changes in the economic policy, this traditional system, which, apart from producing fish, produced 100–140 kilograms of shrimp per hectare of land, gave way to a more intensive shrimp mono-culture which could produce thousands of kilograms per crop. A number of private companies and multinational corporations, including the Indian Tobacco Company, Tata, Hindustan Lever, and the Thai aquaculture giant Chareon Pokaphand, started investing in shrimp farms. All these companies are engaged in intensive farming to maximize short-term profits.

The real boom came in the late 1990s and resulted in the establishment of shrimp farms on a war footing. The Union Ministry of Food Processing Industries gave a glimpse of the situation in this sector: A number of big and small companies were tying up with international majors in a bid to make the best of opportunities in shrimp farming. DCL Maritech, the Andhra-based group, was tying up with CP Agriculture, a Thailand-based business group to set up an intensive shrimp unit in Andhra Pradesh. This 100 per cent export-oriented unit (EOU) was expected to clock a Rs 39 crore turnover by 1994–5. Two other shrimp projects which came up in Andhra Pradesh were: Tirumala Fukitech Aquafarms (Rs 11.90 crore) and Auriferom (Rs 9.50 crore). The US-based Consolidated Sea Food Corporation had plans to float a $275 million project. Kavini Fisheries collaborated with Asia Pacific Seafoods and Monotech Singapore to set up a 100 per cent EOU at Rs 48.6 crore investment in Nellore in Andhra Pradesh; Southern Sea Foods planned to have tie-ups with Wimpy's, UK and Mitsubishi, Japan in Kerala and Tamil Nadu. The VGP Aquafarm launched an export-oriented shrimp farm in an area of 1000 acres in the coastal areas of Tamil Nadu. Many others, similarly, cast their nets wide and deep.

The lure of quick profit prompted the development of large-scale shrimp aquaculture projects. Consider the select financial data of six important aquaculture companies in the country compiled by the Ministry of Finance, i.e., Alsa Marine, Fishing Falcon, Innovative Marine Foods, Kings International, Mac Industries and Rank Aqua Estate. The sales revenue of these companies soared by 176.1 per cent to Rs 83.8 crore (Rs 30.3 crore) in the first half of 1993–4. Rank Aqua Estate increased its sales revenue by 589.8 per cent to Rs 12.2 crore (Rs 1.8 crore); Innovative Marine Foods by 404.3 per cent to Rs 7.1 crore (Rs 1.4 crore); Mac Industries to Rs 26.3 crore (Rs 8.2 crore); and Alsa Marine to Rs 25.3 crore (Rs 19.0 crore) in the first half. As expected, the gross profits of these companies registered a quantum jump of 274.5 per cent to Rs 9.1 crore (Rs 2.4 crore). The net profit nearly tripled by 269.2 per cent to Rs 6.8 crore (Rs 1.9 crore).

Artificial feeds, intensive energy for pumping water and intensive water use are the essential preconditions for semi-intensive and intensive shrimp farms with stocking rates of 100,000 to 300,000 shrimps per hectare. It is critical to maintain water quality, salinity, temperature, and oxygen level, so that intensive stocking can take place and the pollution caused by excessive feeds, faeces, and other organic wastes balanced. An intensive pond requires the regular pumping of seawater of 30 to 35 ppt. salinity mixed with pumped groundwater to keep the 15–20 ppt. range required. It is estimated that roughly 6600 cubic metres of fresh water is needed to dilute full seawater in a one hectare pond at one metre water depth over a cropping period of four months. Thus, the large area of land, big tanks, giant pumps, heavy doses of antibiotics, and fertilizers require a highly capital- and technology-intensive project, to which the local people have no access or relevance. As the shrimp farms spread, there is a destruction of agricultural livelihood and food production, destruction of land, forests and marine fish stock, salinization of groundwater, pollution of the sea and coastal agriculture, displacement of fishing communities, drinking water crisis, social conflict, and violence.

In Tamil Nadu, Nagai-Quai-de Millet, Ramanathapuram and Chidambaranar districts witnessed a boom in shrimp farming. Nagai-Quai-de-Millet district forms part of the old Thanjavur district, known as the 'granary' of Tamil Nadu due to the fertility of the soil. All the cultivable land here is under wet cultivation due to the availability of water resources from the river Cauvery and its distributaries and the relatively well-planned irrigation system. A survey conducted by Nagai-Karai District Fisher People's Forum showed that more than 15,000 acres of cultivable land had been acquired by various companies for artificial shrimp culture in this district alone. There were about ten big and ninety small companies which had acquired this land.

Pudukuppam is a fishing hamlet 15 kilometres from Roompukar, a historic place usually referred to as Keveripoompattinam, at the roots of river Cauveri. The 175 families living in this village are traditionally involved in fishing in the Bay of Bengal. The Neythavasal canal, an irrigation canal of the river Cauvery flowing adjacent to the village, was the only source of survival during the off-seasons, as this canal was abundant with rich varieties of fish. Sri Ram Marine Harvests Ltd set up its farm on 300 acres of land near this village. After harvesting two yields of shrimp the company was in the process of expanding its area of operation to another hundred acres. The newly extended area lay just 50 metres away from the seacoast during the low-tide season.

The villagers alleged that the parent company initially appropriated more than twenty-five acres of village common land (*poramboke* land), including the pathway of the fisherfolk along the coast. About three hundred trees of coconut, palmyrah, etc. had been removed from the poramboke area to construct tanks, walls, and a pumping station. The seawater was pumped into the tank with two motors of 90 horsepower each. The sewage of the farm was let into the sea through three artificial canals (25-inch width and at 4-inch depth each). One sewage canal mixes with the Neydalvasal irrigation canal. Two new companies, Prawnex Sea Food and Krishna Farm, had also acquired a hundred acres of cultivable land in the same area.

Arumugam, a villager, complained, 'The company's fencing blocks our access to the sea. The new structures are causing intensive sea erosion. The effluents stink to high heavens. We have no place to keep our catamarans. The pipeline construction damages our fishing nets. Still the company is hunting for more land to expand.' A village woman said that the underground water had become saline, compelling the village women to walk 1.5 kilometres to fetch drinking water. Women fish vendors/headloaders were restrained from using the traditional pathway along the coast and now they have to trudge 7 kilometres to reach the market.

In Vanagiri village, home to 75 households of fisherfolk and dalits, the Spencer company purchased 350 acres of fertile land. When the company started bulldozing the common pond of the village, depriving them of common water sources, the villagers successfully resisted. A village woman, Sundari, said that in their zeal for constructing structures around the sea, the companies have uprooted trees and bulldozed lands. Kanakamma, another villager, expressed her fear, 'We are landless agricultural labourers. If the company takes away all the agricultural lands, we shall be jobless. If the farm comes, we may have to leave.'

In village after village in the Nagai-Quai-de Millet district, large shrimp farms are coming up. In Sembodai Vedaranyam, Kailimedu Vedaranyam, TRPatnam, Madathukuppam, Vanagiri, Perunthottam, Thennampattinam, Sammankadu, Khargasthu and other villages, fisherfolk, agricultural workers, and the villagers are raising their voices against the shrimp farms taking over cultivable or poramboke land.

In Thalampatti, a fishing village, Abirami and Parisa companies purchased coastal lands and took up almost the entire coast for shrimp culture. When the companies started cutting trees, the villagers successfully resisted it. In Nayakkarkuppan, another fishing village, Magna Foods and Protein established a farm on a hundred acres of cultivable land. The villagers have so far successfully prevented any construction work by the company. In Pudukuppam village, fisherfolk prevented the Sri Ram Marine Harvest from digging a fourth sewage canal. Perunthottam, Chinnoorpettai, and Mayiladuthurai villagers are resisting the setting up of shrimp farms through agitational action. In Tennampattinam, when the landless scheduled caste villagers opposed the new shrimp farm, henchmen of the farm owners burnt down 34 houses of dalit families. Some activists, who sustained serious injuries, were kept in police custody without medical attention.

Against this background, 64 fishing villages of the district collectively chalked out an action plan against the shrimp farms with the slogan 'Stop the sale of land, stop the seed, stop the water.' P. Christy of Nagai–Karai Districts Fisher People's Forum and Chandra Mohan of Workers Peasants Movement explain that now the villagers will stop any construction activity related to shrimp farming through direct action. They will stop the sale or transfer of land for the farm. And current farms will not be allowed production from the next year.

S. Jagannathan, the 82–year-old Gandhian leader of The Tamil Nadu Gram Swaraj Movement and Land for Tillers Freedom, who led the agitation, warned that in the coming days thousands of acres of land already acquired by farm companies would be seized and farmers would start cultivating them. Land for Tillers had launched a one-year Gram Swarajya Padyatra in the district. During its course, having seen the serious problems posed by shrimp farming, they halted their march to organize the people against it. Shrimp companies had enticed big landowners by offering very high rates. They violated the law by taking over cultivable land, which had yielded two crops of paddy and one crop of cotton, groundnut or pulse. They grabbed poramboke (village, common or government land) with impunity. Government attempts to stop this had been stonewalled by court injunctions.

The coastal areas of Andhra Pradesh are also facing the same crisis. Noted environmentalist and activist Vandana Shiva, who made an environmental impact assessment, based on field trips to aqua farms in Andhra Pradesh, noted that the first impact of shrimp farming is the destruction of land and forests in the coastal region. The shelter-belts of casurina, prosopis, and palmyra were cut to make pumping stations, aqueducts and fishponds. In Kurru village, Nellore district, there is no drinking water for approximately six hundred fisherfolk owing to salinization of groundwater. Following the protests of the local women, drinking water was supplied in tankers. Increasing salinity of groundwater has destroyed paddy fields. The environmental impact of fish farming has depleted massive fish resources. In the fishing village of Ramachandrapuram, after Rank Aqua and Siraga Farms commenced operation, the fishermen's shrimp catch came down from about Rs 50,000 per catamaran per month to about Rs 5000 within one year. They used to grow enough ragi for themselves. The 'doruvu', the small ponds for irrigating ragi, having become saline; there is no ragi production any more.

According to O. Fernandes, Director, Human Rights Foundation, among the various visible consequences of shrimp farming and aquaculture in the area, are large-scale purchase and conversion of fertile agricultural lands and illegal utilization of poromboke land for shrimp farms, use of mangroves and brackish water lakes for shrimp farming, collection of brooder prawns and prawn seeds from the sea, estuaries and other water bodies, large-scale unemployment of landless labourers and alienation of lands of small and marginal farmers, the pollution of canals and water bodies because of the heavy use of chemicals and steroids, etc. In numerous villages, the highly toxic effluents being discharged into the canals and seas have also given rise to new diseases, especially skin diseases.

In September 1994, at a national consultation on Aquaculture and Challenges to Human Rights, organized at Chennai by Sneha, a voluntary organization, and the Tamil Nadu Environment Council and the Human Rights Foundation, the participating organizations, trade unions, and activists working along the coastline of Tamil Nadu, Andhra Pradesh and Pondicherry, highlighted the various adverse effects of shrimp aquaculture. The meeting demanded a halt to all shrimp aquaculture farming along the coastline of Tamil Nadu, Andhra Pradesh and Pondicherry, prohibition on the use of agricultural lands, wastelands or coastlands, mangroves, brackish water and salt pans for shrimp farming and prohibition on the collection of brooder prawn and prawn seeds from the sea, estuaries and other water bodies.

On the other hand, a spokesperson of Aquaculture Foundation of India (AFI) explained that chemical industries, heavy metal industries and agricultural pesticide residues are far more dangerous than the bio-degradable waste of aquaculture industry. With proper management, it is possible to recycle such waste and protect the environment. Problems such as salt water seepage into the agricultural areas and the discharge of untreated effluvium into open water bodies could be taken care of. The aquaculture industry could create direct employment for at least two million people and indirect employment for at least twice this number in the coastal rural areas. The skill levels required are minimal. Several ancillary industries could also come up to support aquaculture. V.J. Chandran, Vice-President, Prawnex and a leader of the Aquaculturalists' Association of Mayiladuthurai and Nagapattinam in Tamil Nadu, also explains that in agriculture, a ninety-day crop provides work only for ten days. Aquaculture, on the other hand, provides year-round work at double the rates. Processing and packaging provides additional work, especially for women. Chandran believed that the movements against shrimp farms in Tamil Nadu were contemplating bloodshed and the villagers who stood to gain from the growth of prawn farming would fight back. R. Kothandaraman, Chief Executive Officer for Spencer's aquaculture argued that enough care was being taken to ensure that there was no seepage of sea water into the neighbouring lands and the talk of water turning saline was exaggerated. Since no worthwhile agricultural operation had been undertaken in the area for the last twenty years, the argument that cultivable lands were being converted into shrimp farms was also misleading.

Shaji R. John, Chairman, Kings Group, and Managing Director, Kings International Aqua Marine Exports presented a different perspective. According to him, in India, deep-sea fishing was developed only in Visakhapatnam, but all the trawlers were going in for shrimps. At the current rate of exploitation, the entire shrimp stock would be depleted within ten years. Therefore, the emphasis was now shifting from marine fisheries to aquaculture or land-based fisheries. Before the advent of aquaculture, India held the premier position in fishing, but during the last five years, the situation had changed. Other countries like China, Indonesia, and Thailand, had developed their entire available land and were now ahead of India. Deplorably, India had not developed even 5 per cent of the available land of about 1.4 million hectares. Half of this available land, or even 20 per cent of it, was capable of producing staggering quantities of fish products. But this could not be achieved by a small number of companies. Instead, it had to be done by the local people. The big companies could only serve as model agencies in particular areas, taking up franchise farming and disseminating technology to the local farmers, buying back their produce and also helping market produce in the international market. Shaji conceded that most of the companies entering aquaculture were lured by large profits and did not know the local place, the people, and their language. Hence, the local people viewed them as outsiders encroaching upon their areas. The Chilka project of the Tatas had to be shelved because of these problems. As against the existing local price of Rs 60–70 per kg for the prawns, the Tatas were prepared to pay an international price of Rs 250 per kg. But the most affected middlemen created several problems.

Governments, though slowly realizing the threat of the haphazard growth of shrimp farms, are postponing any concrete action and restricting themselves to commissioning studies and constituting committees. First, the Union Ministry of Environment and Forest initiated a project to study the impact of the effluents discharged from aquaculture ponds, stretching over three years and covering the shrimp farms from Nellore on the Andhra Pradesh coast, up to the Kanyakumari coast in Tamil Nadu. Thereafter, the Ministry of Agriculture constituted an expert committee.

The Tamil Nadu government appointed an expert committee to

suggest norms and guidelines for a legislation to regulate the growing aquaculture industry, particularly shrimp farming. Among the objectives of the legislation were to ensure that shrimp farming did not pose a danger to agricultural land and sub-soil water, that groundwater was not tapped for the industry and to fix pollution standards on effluents discharged. The proposed legislation also sought to impose restrictions on the conversion of agricultural land for shrimp farming.

Latika D. Padalkar, former Commissioner of Fisheries, Government of Tamil Nadu, proposed guidelines for the aquaculture industry in the state, 'It seems desirable to create zones where aquaculture industry can be set up. Leasing of poramboke lands for aquaculture *per se* should not be done in certain regions. Since most parts of coastal areas have very small columns of potable water at shallow depth, the industry may not use deep aquifers. Effluents need to be collected and treated before being let out, at the time of their final outcome from the farm. The State Pollution Control Board should come into the picture. They should take samples of effluents generated by industries and enforce conformity of these standards as per the Water Act.'

A review team of the committee toured Thanjavur district and listed the complaints of villagers and traditional fisherfolk against the shrimp farms. The committee planned to finalize its regulatory scheme by December 1991–3. A senior official of the Department of Fisheries, who was a member of the review team, pointed out that implementation of regulatory and environmental safety laws was not easy, as an estimated 40 per cent of the shrimp farms in the state, which were major contributors to the Rs 320 crore of shrimp export from Tamil Nadu, are unregistered. The Andhra Pradesh government also constituted an expert committee and proposed a draft legislation to regulate the shrimp farms and aquaculture industries in the state. The committee assessed that over one lakh acres of land had been converted into shrimp farms along the 974 km Andhra Pradesh coastline stretching from Ichapuram to Tada, particularly in the districts of Godavari, Krishna, Nellore, Visakhapatnam and Vizianagaram.

The shrimp aquaculture industry now is a multi-crore industry in India and it is widely believed that the industry will continue to grow. The poor and oppressed pay a high environmental and social cost for this expansion, which is being largely disregarded by governments and the companies heavily involved in intensive shrimp culture. The struggle against it, however, is also bound to continue.

Coastal Regulation Zones
The Pauper, the Thief and the Convict

In the coastal zones of India, violation of the environmental regulations seems to be the norm. The National Fishworkers' Forum, a federation of trade unions among the coastal fishworkers, has done a comprehensive exercise of mapping the extent of Coastal Regulation Zones (CRZ) violations on the Indian coast. The report on this exercise points out that the worst violation of India's coastal environment are tourism, industry, infrastructure, reclamation, aquaculture, and mining. The forum called for a fisheries strike on 10 May 1999, to press for the implementation of the CRZ Notification, and for action against its violators.

India has a coastline of 7500 kilometres, spread over nine coastal states. India is officially claimed to be the only country in the world to have brought about CRZ Notification to protect the coasts. The notification, introduced on 19 February 1991, initially directed that the beaches had to be kept clear of all activities for at least 500 metres from the high water line (HTL). However, sustained pressure from the tourism and shrimp farming industries saw to it that the distance was reduced, first to 200 metres and finally to nothing. This happened after a case was brought before the Supreme Court which, through a landmark judgement on 18 April 1996, reinstated the original position of 500 metres, and directed the formulation of Coastal Zone Management (CZM) Plans in the coastal states. The court further directed the demolition of all prawn farms set up within 500 metres of the HTL and alongside creeks, backwaters, estuaries, rivers, etc.

Harekrishna Debnath, Chairperson of the Forum, explained that the judgement of the Supreme Court showed how neither the coastal state governments nor Government of India took steps to implement

Published in *Inter Press Service*, 4 January 1998.

the CRZ Notification. Nevertheless, when the court directed the states to attend to the matter, the Forum also moved into action, to ensure that the CZM plans were objectively drawn up. When the submitted plans were disregarded by state governments and there was a move to scrap the judgement and the Notification in Parliament, the Forum felt the need to create wider awareness on this issue. It decided to map the ongoing violations in the coastal zones.

Environmental groups, people's movements, and civil liberties fora also raised an alarm over the serious violation of coastal regulations. Bittu Sahgal, an environmentalist, and a member of the Coastal Task Force, constituted by the Ministry of Environment and Forests, in a communication to the Ministry, noted that land developers around the country worked overtime in tandem with state government officials to cash in on thousands of crores of rupees worth of public lands. A similar move on forestlands and coastal belts would radically alter the survival ecology of over 300 million people within three to five years. He foresaw that the coastal areas, the breeding grounds of fish upon which coastal communities survived, would be replaced by prawn farms, five-star hotels, thermal plants, chemical and petroleum complexes, copper smelters, coastal highways and the urban sprawl.

The survey by the National Fisherworkers' Forum gave a state-wise analysis of CRZ violations. Ten violations were noted in Porbander, Jamnagar, Bhavnagar, Kutch, Bharuch, Daman and Valsad areas of Gujarat, by the shipbreaking and cement industries, while the rest were on account of infrastructure development activities, like construction of jetties and ports. In Maharashtra, forty-four violations were recorded in Raigad, Mumbai, Sindhudurg, Thane and Ratnagiri districts. Reclamation and construction-related activities together accounted for 69.45 per cent and industrial activities for 28.81 per cent of the violations. In the Karnataka coast, of the thirty violations, 82.76 per cent were tourism related, and concentred in Udupi, Mavalli and the neighbouring areas.

In Kerala, violation mapping was carried out throughout the coasts. Though Kerala had a well-prepared CZM Plan, the report alleged that the state government intended to relax the notification, so as to accommodate 'development' in the state. Different committees had been constituted to dilute the accepted plan. Of the 377 violations recorded in the state, a major chunk, 39.59 per cent, was on account of industrial activities, which include large projects like the naphtha-based power projects at Kannur and Vypeen, and small-scale units.

Tourism accounted for about 28.12 per cent of the total, and of 106 violations in this category, 73 were new constructions. Similarly, all the 21 infrastructure items, which contribute to 5.77 per cent, were newly mapped.

In undertaking the mapping exercise, the survey team of 200 youth volunteers, led by Dr D. Nandhakumar of the Department of Geography, Kerala University, made the official CRZ document its basic working document. Procuring such documents in Maharashtra, Tamil Nadu, Daman and Diu, and West Bengal became a problem by itself. In Gujarat, the CZM Plan submitted by the state government correctly listed some mangrove forests in Kutch as part of the CRZ. However, presumably to suit the Sanghi Cement Company, which wanted to build a jetty in the fish-breeding area, the government altered the maps. In Maharashtra, vast stretches of mangrove forests were completely left out of the maps submitted. In Karnataka, some major tourism projects proposed, for example, the Tannirbavi Project on 124 acres of land, which had to be handed over by the Karnataka Industrial Area Development Board to the state tourism department, were not mentioned in the CZMP. In Kerala, the areas east of the Thannermukkam barrage were excluded from the plan. The Kumarakom, which has an extensive stretch of mangroves, was excluded on the grounds that the barrage was closed for more than six months of the year. The Kumarakom mangrove stretch is a well-known bird sanctuary, and the Kerala Shastra Sahitya Parishad, a people's science group, has filed a writ petition in the High Court against this omission.

In a tour of the aquaculture areas of Andhra Pradesh, V. Vivekanandan of the South Indian Federation of Fishermen's Cooperatives discovered oddities in the implementation of CZMP. Ensuring compliance with the Supreme Court's order was the responsibility of the district administration, specifically the District Collector and the Superintendent of Police. But confusion reigned. In West Godavari and East Godavari districts, officials in the local administration seemed to be under the impression that the judicially established limit was 50 metres or 100 metres, and not 500 metres. Thus most of the aquafarms were not excessively bothered.

The report of the National Fisherworkers' Forum also made several suggestions for the improvement of the CZMP—the establishment of a Coastal Zone Management Authority in each coastal state, clearly defined penalties/punishments for violations, special consideration to the customary rights of the coastal communities, focus on an Area Specific Management Plan, and right to information on CRZ for coastal communities and local bodies. The forum had its own campaign plans. As its chairperson outlined, the report and the team of volunteers who made it possible, would act as a catalyst in their campaign all over the country on these flagrant violations. They would also continue to monitor the breach of regulation. There would also be recourse to legal action, and filing of contempt petitions in courts. There was also to be another strike at the sea to force governments to implement the CRZ Notification.

The Cursed Dam

Baba Amte and Medha Patkar, Badwani, MP

*Protest against Sardar Sarovar Project—Sixty villages
come together at Manibeli*

Sardar Sarovar Project
Life in an Emergency

'It has come! It has come!', exclaimed an excited group of villagers of Badwani in Madhya Pradesh. 'What has come?' I asked one of them curiously. Answered all, almost in unison, 'Our *Lok Nivada*'. They kept repeating it. I moved to another group of villagers and they too said the same.

Lok Nivada was a people's referendum in the villages affected by the Sardar Sarovar Project (SSP) spread over the states of Madhya Pradesh, Maharashtra, and Gujarat. It was a remarkable campaign launched by the *Narmada Bachao Andolan* (NBA) on 30 January 1993 in Manibeli (Maharashtra) and Vadgam (Gujarat), the first villages to be submerged. Held in every project-affected village on either side of the Narmada, it managed to elicit opinion on the project from all the families inhabiting the area. The detailed questionnaire prepared by the NBA was filled in and signed by thousands of villagers, with women in tribal areas being consulted independently, and their families being included in the count only with their approval. The consultation with women members of the family was made mandatory so that their views too got equal preference.

According to the results of the Lok Nivada, 22,523 families opposed the construction of the dam. More than two-thirds of these families lived in the submergence zone of the dam. The writing on the wall was clear. A signed declaration said, 'We, the villagers, have now understood that the Sardar Sarovar Dam is a destructive project in human, economic and environmental terms. Therefore we resolve that we will never leave our forest, water, home, land, *mandir*, *masjid* and mother Narmada at any cost.' In keeping with this resolve,

This is a substantially revised version of the paper originally published in *Frontline*, 21 May 1993.

village houses displayed the slogan, 'We may be drowned, but we will not move out'.

The results of this referendum were sent to the central and state governments, the World Bank, and countries involved in the project. 'Anybody can verify the validity of our survey', said activists of the NBA. The logic behind the campaign, explained NBA leader Medha Patkar, was to reassess public opinion about the dam. Do people want resettlement? Do they want the movement to continue? Would they finally come to Manibeli, the first village to be submerged, at the critical juncture? On all these questions, she felt, the villagers responded honestly on the whole, with some families expressing their preference for resettlement.

As the conflict heightened, all the parties stepped up the preparations to defend their respective positions. Dam construction was in full swing. B.J. Parmar, Chief Engineer and in-charge of dam construction, said in Baroda (Gujarat), 'By the monsoons, the height of the dam will reach 61 metres and some areas will be submerged for the first time, depending on the amount of rain water we get this year.'

In Maharashtra, Manibeli and a number of other villages were issued notices that water and submergence would affect them in July 1993 and they had to vacate their villages by 31 January. In Gujarat, the state government filed a writ petition in the high court in January, for vacating the stay, by the same court, on the displacement of the people of Vadgam, the first village to be submerged. In Madhya Pradesh, a survey of the properties of those families who were living in Alirajpur tehsil of Jhabua district which was about to be submerged, started earlier that year. In a confrontation, which lasted for many days, the police resorted to firing twice in Anjanwara village. Vinod Kumar, SDM, Alirajpur tehsil, admits, 'These are final, emergency hours. Villages will be drowned within six months. No other way is left. We have to remove people. We have completed the property survey of all the 26 villages likely to be affected.'

Kesubhai in Manibeli, Bhulabhai in Vadgam, and Bawa in Jalsindhi village hail from Maharashtra, Gujarat and Madhya Pradesh respectively. Nevertheless, they had two things in common—they would be the first to go under the waters of Narmada, when the rains came, and they refused steadfastly to be resettled. In the emerging standoff, they were the symbols of protest.

The Narmada Development Authority's publication *Narmada Vikas Varta* (Issue No. 2, August 1992), in a piece entitled 'Current Situation in Manibeli', said, 'Out of the total 222 PAP (project-affected persons) families, 76 would have been affected by the 'one in hundred year flood' level of the monsoon of 1992. However, 75 of these families have shifted to the resettlement sites and the only remaining hut, i.e. of Kesu Dhedya, was shifted to a higher level in the village.'

In reality, Kesubhai's hut remained at its original spot. Above it proudly flew the flag of the NBA: 'Save Narmada, Save Humankind.' Kesubhai owned nine acres of agricultural land in Manibeli. His wife Ganga said that the family took a joint decision to stay on. Their daughter Mangi worked actively with the NBA and was even jailed. They defiantly asked, 'This is our house, our land. Why should we leave it?' They also said that Kesubhai's elder brother had moved to a resettlement colony in Parveta, Gujarat, and three of his family members died there from 'mysterious' diseases.

In Gujarat, the high court's stay on the eviction of Bhulabhai from the lowest level in village Vadgam was still valid. The Narmada flowed close to his hut. Said Bhulabhai: 'I have five sons and all of them have grown up here. I cannot imagine shifting from here. We do not want to die again and again after being displaced and resettled. It is better to be drowned and die once. That is why I have refused the resettlement offer.' His son Bankhad affirmed, 'It is not a question of our house alone. Our farming, our livelihood, our life—all these are involved.'

Director of Rehabilitation (Sardar Sarovar Narmada Nigam Limited), Rajagopalan, contended that Bhulabhai does not have PAP status because he settled in the village after the cut-off date. However, he could appeal against the tribunal's decision.

Of the thirty households in Jalsindhi, the first village to be submerged in Madhya Pradesh state, eight were resettled in Gujarat. But the rest refused to move. Said Baba, 'The rest of us twenty-two families will not move anywhere. We did not go anywhere to see the resettlement area.' They argued that the land in Jalsindhi is fertile, and fuel and fodder were easily available.

4 March, Badwani. This was where Mahatma Gandhi's ashes were immersed in the Narmada forty-five years ago. More than ten thousand villagers waited here to greet the jathas of villagers and activists, to mark the conclusion of Nivada. Tribals, dressed in traditional costumes, sang and danced and others performed plays and raised slogans of solidarity. Later in the evening, their voices joined that of Baba Amte, also leading the campaign, to vow, 'We will not leave.'

In an interview, Baba Amte said, 'We are building our dam of love, cooperation and belief to confront the government dam. If the government is willing to talk, we are ready on the basis of people's participation. However, the government should first realize that it cannot continue with the project at gunpoint. We are pursuing a peaceful path, but it is also the responsibility of the government that peace is maintained in the Narmada valley. The manner in which it is trying to crush us, this peace may not continue for long.' The voice of the famous Gandhian was choked with pain and anger.

The firebrand Janata Dal leader Mrinal Gore, addressing the concluding meeting, said that it was proven that the Sardar Sarovar Project would not solve the water problems of the Kutch–Saurashtra region of Gujarat. They had to ask themselves a more fundamental question: Should a poor country like ours continue with these heavy capital-based projects under the diktats of the World Bank and developed countries?

The last programme of the Nivada was held at Manibeli on 5 March, where representatives of more than sixty villages made a bonfire (holi) and fixed tags on the trunk of a tree, signifying their unity and resolve. Each tag bore the name of the village and the trunk was almost covered with them.

Here we met Gopichand, Hiralal Jaylal, Ranchod Jaylal, Parsingh Rana, Kakaria Guman, and Sadia Singh Lonia from Amlali village of Madhya Pradesh. They had willingly resettled in the Gajadhara village near Baroda, but returned to their village in just eight months. They complained that the land allotted to them was fallow, and they had no access to fodder and fuel. Now they resolved to resist resettlement. Nathu and Nataro were joining them from Julkhera village of Baroda district, Gujarat. They had also returned to their village from a resettlement colony because 'there were no facilities there'. Although Nathu's land did not fall within the submergence zone, it would be surrounded by water on all sides. Yet he received no compensation. Thus, he came here to join the movement.

Parweta is a known resettlement colony in Gujarat, where displaced people from the three states have been settled. The colony was strictly divided into three parts, corresponding to the native states of the resettled people. Hira Bhai Mansukh Bhai came to here two years ago from Junugaon, Vadgam, Gujarat, but he has got no land till now, and he complained that there was no electricity or water in the colony. The only well in the colony is a constant source of quarrel. Maniben Hiraben who is from Madhya Pradesh, complained that her husband did not seek her consent about the resettlement. If the Sardar Sarovar Project were to fall through, she would be among the first to return to her village, Manibeli. As Dhananath, also from Manibeli, said, 'In the resettlement colony, there is always pain and anguish, and no livelihood. It is better to be drowned there.'

Questions of conscience notwithstanding, the fact remained that a majority of the villagers refused to budge even in an emergency. Will the government leave them to be drowned? Will it be easy for the World Bank to decide in favour of continuing with the project at their crucial meeting on 31 March 1993? Indeed, there were difficult choices ahead.

Suvarnarekha Project
One Night and Now What?

On the night of 28 August 1991, in Chaibasa District of South Bihar, fifty-two villages submerged unexpectedly. Hundreds of people, encircled by water from all sides, searched for safety in the dark. Their houses, crops and other properties were washed away. Many climbed trees, crying out for help.

The submergence was caused neither by heavy rains nor by a flood. It was part of the destruction wrought by the Chandil dam. When, for the first time, the catchment area of the Chandil dam was being filled with water in the ongoing monsoon, it created havoc. Villages and lands which were declared safe according to the dam's plans were threatened with submergence. Places whose submergence was due two to four years hence, at later stages of the construction, were already inundated. The estimation of the extent of submergence on account of the volume of water in the dam went beyond calculation.

Chandil dam comes within the Suvarnarekha multipurpose project, which is a composite project of two dams, two barrages and seven canals. One dam is being constructed on the Suvarnarekha river near Chandil (Jamshedpur). The other is on the Kharkai river near Icha (Chaibasa). The project cost at the time of its initiation in 1974 was Rs 129 crores. By 1991 it had gone up to Rs 1500 crore. According to the plan, the dam and its reservoirs will displace 194 villages and submerge 38,587 hectares of land.

So far, construction up to 184 metres height of the dam has been completed, against the maximum height of 196 metres. However, the monsoon and the subsequent unscheduled submergence of a large number of villages raised serious questions regarding the so-called

This chapter is a substantially revised version of the collection of articles that was originally published in the *Navbharat Times*, 24–7 October 1991.

scientific assessments on submergence, water level and catchment area of the dam, and the life and security of the villagers. Since the first phase itself had gone awry, what could happen in the second and the third is best left to pessimistic speculation.

Two persons, including a pregnant woman were dead and nearly 30,000 families were affected. Thousands were trapped in the water-logged areas. With despair and disease everywhere, uncertainty and tension reigned here. The villagers lost much, and they only earned the pitiable epithet *Jaldubi parivar*—dispossessed families.

Upendra Sinha, Officer In-Charge of Flood Control Cell, Chandil Dam, gave a detailed official account of what happened since 26 August: 'The project had the premonition, based on their technical survey, that at the level of 177 metres water in the dam, in this year's monsoon, only fourteen villages would be submerged. On the night of 26–27 August, the water in the dam reservoir started increasing due to the rainfall in the catchment area. On 28 August, water reached a level of 177 metres. Due to the submergence, six villages with 399 families have been fully affected. Their agricultural land and houses have been completely innundated. These villages are Borabinda, Gopalpur, Harsundarpur, Kesar-gadia, Birdi, and Durrie. Twelve villages have been 'particularly' affected, with agricultural lands to-tally under water. There is another list of 29 villages, which are declared partially submerged. According to our estimate, the submer-gence damaged Rs 50 lakh worth of houses, crops and other valuables in the village.'

Dam projects in this country are indifferent to the fate of the poor and the tribals who are displaced. Of course they promise many things. But from the inception to the completion, these promises are broken or changed to suit vested interests. However, can a dam be so cruel and insensitive to the lives of the local population? Can the dam authorities be so casual about the impending threat to hundreds of lives? Can they afford such a miscalculation about the measurement of height and breadth, and its disastrous impact?

The confidential status paper on submergence prepared by the dam authorities can have serious implications for the dam. A village like Odiya, located at a height of 182 metres as per official records, was submerged when the water level was at a height of only 177 metres. Villages like Bandu and Dalgram, situated at a height of 185 metres, also submerged at 177 metres of water. These villages were supposed to be submerged in phases two and three, within a span of the next two years. But tragedy befell them without warning in the first phase itself. Janum village was not notified among the villages but it too was submerged. Village Durrie, included in the list of partially affected villages at 180 to 182 metres, was totally engulfed by water.

According to Upendra Sinha, instead of carrying out surveys at the field level, to measure the height of the villages, the Investigation Department relied on secondary information from the surveys carried out by the Survey of India and other agencies. Since the submergence area was mainly an undulating area (and a village consists of various *tolas* at different heights), such studies should have been carried out long before the construction of the dam had started, he said.

According to the Superintending Engineer of Chandil Dam, J.N. Ganju, a report had been sent, prior to the monsoon, to the Chief Engineer and the Project Administrator that the water could reach 182.6 metres, and heavy loss of both human lives and property could take place. The report also expressed the apprehension that 177 metres of water in the dam could affect at least 45 villages. But the project authorities decided to ignore the report. Also, rehabilitation work is not complete in a large number of villages, and the acceptance that forty-five villages are going to be affected has jeopardized the future of the project to some extent.

From village Durrie, a seventy-member delegation had met the Project Administration to assure themselves that their village was safe from submergence. The Administrator assured them that Durrie did not come in the first phase of submergence. On his assurance, they started cultivating their agricultural lands. Within days, they lost their houses and their agricultural land forever. The Project Administrator, Devdas Chote Roy, denied knowledge of any report regarding fears of submergence but was candid that something had seriously gone wrong, especially regarding the submergence area of the dam. He blamed his junior technical colleagues for not assessing the height accurately.

The district administration, which had virtually no say in the project, had to bear the responsibility for providing relief for the affected families. Rajbala Verma, District Magistrate, said in Chaibasa, 'Such happenings are not acts of nature, but man-made disasters. Serious lapses have taken place this time and this prompts rethinking on issues of survey, water, catchment areas, height of the dam, etc.' The project authorities had sent them inaccurate estimates of submergence and water level, she said, and expressed the view that unless there was a new survey, the sluice gate of the dam should not be allowed to be closed in the following year, nor work on the dam allowed to go on. Otherwise, by the end of the following year the height of the dam would increase to 186 metres, which would mean more danger, more water and more submergence.

✛

In a relief camp in the Ichagarh school building, near the Ichagarh police station, hundreds of villagers were cramped in classrooms and verandas. A Block Development Officer had been held captive there for a few hours by the agitated villagers, who demanded that the government officials must go and see for themselves the damage being inflicted in the submerged areas. Many had reached the relief camp after crossing the river, and walking for over four hundred kilometres.

Complaints about inadequate relief work were common. Desperate, destitute, and angry they all had unpleasant things to say about the administration. 'No compensation for the loss of standing crops, livestock, household items, and houses submerged is being offered. The officials say that we are not eligible for any compensation for the loss of properties, because we had been issued Section 4 notice as per the Land Acquisition Act. If the Section 4 notice had been served to us, why is it that we have not been rehabilitated so far?', asked Harihar Poddar.

As relief, they were given Rs 2 for every adult and 3.5 kilograms rice per family, per week. The relief package was limited to two weeks. To get this relief, people had to spend almost a day. The head of the family had to be physically present to receive it. The relief was being distributed on the basis of a *Vikas Pustika*, prepared in 1985.

Medical facilities in the relief camps were meagre. 'Medicines were available at the relief distribution centres, 8–10 kilometres away from the camps. The doctors did not visit the relief camps and distribution centres. Even the village health workers did not visit the camps,' complained Ajit Kumar, displaced from Janum village. At Chilgu village where a relief camp had been set up, there was a Public Health Centre with only one ANM dealing with about 20–25 patients with various complaints every day. To get medicines at this centre, it took two days to secure the approval of the health authorities.

The submerged areas were still under water after fifteen days, and the displaced persons were being told to vacate the relief camps. No resettlement plan was ready.

It would appear, in these inflation-ridden times, that one can construct a home for as little as Rs 400, and buy a tamarind tree for Rs 2.50. In Boxai village of Ichagarh block, Guna Kalindi got Rs 408.46 as compensation for the loss of his house (Khata No. 204, Survey Area No. 504), Fora Dom (Khata No. 169, Survey No. 514), Devi Kalindi (Khata No. 169, Survey No. 502) and Katala Kalindi (Khata No. 39, Survey No. 502), received Rs 857.78, Rs 980.32 and Rs 970.70 respectively for their houses.

Krishna Gop was denied compensation for his two ponds. Sambhu Mahto and Karmu Mahto of Kalyanpur village received no compensation for their wells. In Hurloonga village, one *palas* (Butea frondosa) tree and one tamarind tree were compensated to the princely tune of Rs 6 and Rs 2.50.

The Chandil dam is expected to be fully operational in a year's time. But resettlement in the region is incomplete, haphazard and *ad hoc*. Payment of compensation, identification of resettlement areas, and the actual resettlement are all incomplete. The District Magistrate, Rajbal Verma said, 'Very little resettlement work has been done in the Chandil dam area. There is much truth in the complaints of the villagers. Leave alone proper rehabilitation, I have asked for the completion certificate for each village regarding compensation money, and even that has not been given by the project authorities.'

In December 1990, a grandiose resettlement policy was trotted out to impress the international funding agency. Resettlement was to be completed before displacement. The would-be-displaced were to be given resettlement sites six months ahead of their displacement. The displaced would be given priority in lower-ranking jobs within the project and in government services in the area. All families whose houses and lands fell within the submergence area would be considered displaced. Every adult member of the family would be recognized as displaced. Every displaced and agriculture dependent family would get a minimum of 2 acres land or Rs 1 lakh to purchase the land. Every family would be given Rs 20,000 to rebuild their house, etc.

To give compensation and other benefits, a survey of the would-be-displaced people was undertaken and they were given an identity book. This survey excluded many, especially women and older people. Kangadin *tola* of Aara village would lose its twenty-two houses, but these families were not identified in the survey nor given any card.

In Kumahari, five families were arbitrarily denied the identity card. Belu Gwalin of Kalyanpur village, Karla Mahtani, Guruwani, and Jai Ramni of Dhathidih village, all of them widows, were denied cards.

Haria in Dumari *tola* of Kashipur village said, 'We are thirty-seven families, and none of us has received a single paisa, in spite of sending our application to one and all in the project.' Kumhari's people complained, 'We are not being surveyed yet, what to say about compensation and payment.' In Moisada village, the lands have been measured, but short changed in the records. Lahtan, with 20.8 acres recorded agricultural land, has received compensation for only 12.8 acres of land.

Many people of Durrie village have been engaged in cultivation as their primary livelihood for more than twenty years, but without any *patta*. They have been told that they will not receive any compensation. Alas, no compensation has been given for the common property resources in the village, such as pastures and common lands.

In the past, the project and the dam had aroused strong reactions among the villages. In 1978, it led to the Joiada police firing, in which four persons were killed. Again, there was police firing in 1983, in which two persons were killed. Over the years, the issues have remained the same: dam, land, submergence, displacement, resettlement. To these questions, organizations like the Visthapit Mukti Vahini and member of parliament Ram Tahal Chowdhary, have raised some new issues like the stoppage of construction works, fresh thinking on the design of the dam and an urgent resurvey of the submergence areas.

28 October—Our Independence Day

A natural waterfall in the hills can be dammed to transform life in a tribal village entirely on the initiative of the villagers themselves. Vishunpur block is in a corner of Gumla district, a dry, hilly, plateau region, 123 kilometres north-west of Ranchi. From there village Sato is a further 12 kilometres by a dusty road. A vast tract of agricultural land surrounds the village. At the top of the 200-foot high Sato hill in the village is a perennial stream. At one end of the stream are a big pit and a thick wall to check the flow of water. A wooden gate attached to the wall is connected to hilly stones carved into nalas. All these are the makings of a dam.

Sato, a village of Urao tribals, has two tolas consisting of 70 houses and a population of 500. All houses have some land, with a maximum of 4 to 5 acres. The dam was completed on 28 October 1988, when the water came down from the peak of the hill to our ponds and lands. A dream came true after more than forty years. It was like Independence Day for us,' said Jale, a village elder who had witnessed several unsuccessful efforts to solve the water problems of the village.

According to the block office, the region gets at least 1200 millimetres of rain, most of it between June and September. The soil cannot absorb much of the rainwater. Like most of the other regions of Chotanagpur, the village faced a perennial problem of drinking water. The water from the stream was being discharged into the far off Koel river without being of much use to the villagers. They had time and again petitioned the district administration and the irrigation department that instead of the usual rural development programmes they should give a serious thought to harnessing the available water resources for domestic and irrigation purposes. But the administration considered it impractical to build nalas, dams and irrigation channels over natural streams.

In the 1950s, they initiated some construction at the top of the hills. Dhirja, an old man, said reflectively, 'It was some time after independence that we started constructing a small bund at the top of the hill. We constructed a pond downhill. But they were not complete. We also thought of making nalas at some places. By that time, the agricultural season had arrived, the villagers got involved in farming and left everything in a lurch.'

Another effort commenced in the 1960s, when the whole region was affected by the worst-ever famine the area had witnessed. The stream was their sole source of water. After the famine they met and agreed that every two households would collect one bag of cement and every household would offer its own spades, pickaxes and other implements. Big stones were piled together to dam the stream and the stones were carved or cut at many places to control the course of water. For the first time, some water reached the tank and agricultural lands in the village. But the first rain washed away everything. The stones slipped away and the dam cracked.

They had little expertise in dam construction. Also, they were short of required funds.

In 1986, a non-governmental organization, *Vikas Bharti* came to Sato. 'We were clear that we first wanted to solve our water problem. We wanted to build a dam to tap the water available in the hills. At our first meeting with the people from the organization, we put forward our priority and told them clearly that we did not need any other help. If irrigation could be ensured and agriculture improved, we could well take care of our other problems', said Tana, who took an active part in the course of events that followed the NGO intervention. The villagers volunteered all the physical labour required for the dam. Every house also volunteered to contribute Rs 100 for hiring skilled labourers. Members of at least nine families would do labour everyday. Shramdan would continue for six months in a year, to be suspended during the agricultural season. The village would take care of the boarding and lodging of the skilled workforce associated with blasting, construction, etc.

When construction work on the dam started in early 1987, and continued unhindered for the first six months, it became the talk of the whole region. Sato itself was so surcharged that on some days most of the villagers, not just members of the assigned nine families, involved themselves in the construction work. It also drew the attention of the district administration. The district magistrate came in the early days of the construction and announced a grant of Rs 10,000 for the village.

The culmination, of the beginning that was made on 28 October

1988, came after two years of Shramdan. Most of the agricultural land of the village could now be irrigated. The villagers were capable of harvesting all crops besides seasonal vegetables. They had reason to be proud of their handiwork. 'We look after the dam, tank, canals. Whenever water increases or decreases, we open or close the wooden gate over the dam', said Jale.

The availability of water solved many problems, but created some new ones. The canal network itself was incomplete while the unlined canals led to water seepage on the way. This made the location of some land more advantageous. Even distribution of water among the villagers was a daunting task. But a majority of them felt that the completion of the canal network in due course would solve their problems, since nature had given them enough water to feed everybody.

Travails of Land

Anna Hazare at Ralegan Siddhi

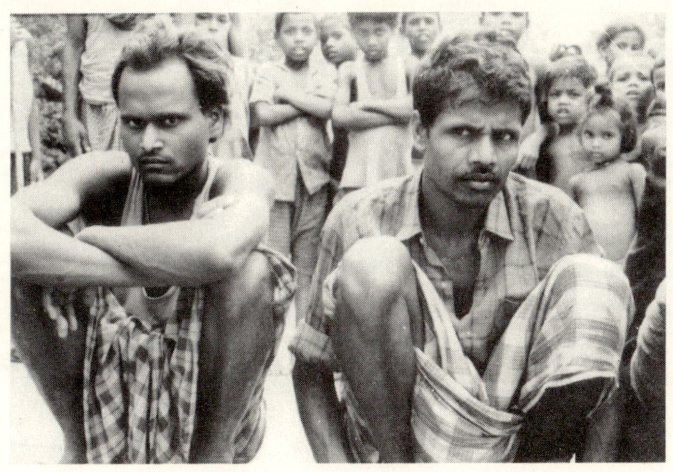

Musahars—undisguised unemployment

Agricultural Land
Land is like Your Best Friend

Over the years land-related disputes have caused numerous conflicts between farmers themselves, and between farmers and forest departments, project authorities, and local administrations. Around twenty thousand cases were pending over land in local and other courts in the state of Uttar Pradesh, Prajapati, a resident of Karahia village of Mirzapur district informed.

Karahia is one of the 433 tribal villages in the Kaimur hill region of southern Mirzapur. The region has a spread of 9.23 lakh acres, and has a population of half a million. The villages are mostly inaccessible. The Kaimur region has never had a land survey, recording or settlement and no land reform measures. Land measurement was begun under the provisions of the Revenue Act in 1964–5 which remains incomplete till date. Out of 9.23 lakh hectares of land, 7.87 lakh hectares, were declared as reserved forests between 1960 and 1970. Five super thermal power stations, four cement factories, one aluminium factory, one carbon black factory, one chemical factory and several lime and coal mines have come up in this region.

When the Rihand dam was constructed in 1960, 80 villages and 1 lakh acres of land were submerged. Approximately 50,000 villagers were displaced, out of whom almost one-third had no land records. They got a maximum of 3.5 bighas land as compensation. For the Rihand Super Thermal Power Station land was acquired in Bijpur, Aghora, Sirsoti, and Dodhar villages. In all, 267 farmers and 550 families were displaced. Of the 267 farmers, 100 were dalits (of whom 80 were tribals) and 167 belonged to backward castes. Since there were no land survey records and settlements in these villages, many

This chapter is a substantially revised version of the collection of articles originally published in the *Navbharat Times*, 8–9 February 1992.

of the dispossessed did not get any compensation. Bijpur village had 3069 bighas of land of which 1133.05 bighas were under private ownership. Only 348 bighas of the latter were being cultivated only by those in whose names the land was recorded. 112.05 bighas were recorded in the name of two farmers. 285.05 bighas were recorded as belonging to people living in other villages, and 387 bighas were on hills, under forest or bushes, not under cultivation, but which were still recorded under some names. The whole land of the village was acquired, but only some were compensated. In villages like Kirweel and Devari, the same plot of land had five or six claimants. Those displaced from the Rihand dam and the landless got pattas for land under the 20-point programme.

In the 1980s the forest department started a massive drive to take possession of land coming within the ambit of reserved forests. The definition of this category was so sweeping that all kinds of land came under the forest department. It included land occupied by a village for 40–50 years, land which was being cultivated by the villagers for a very long time, *gram sabha* land, pastures, and in some cases even graveyards.

Many villagers and tribal families faced eviction by the joint forces of the local and forest administration. Uday Bhan of Myorpur village, Robertsganj tehsil recalled how their houses were razed and the land cordoned off with posts and barbed wire fences by the forest department. When they protested, a number of cases were registered against them under the Forest Conservation Act.

In 1982, at the initiative of Banwasi Seva Ashram, a social organization active in the region since the 1960s, the Bhoomi Haqdari Morcha came into being to help the dispossessed households. The movement focussed on the following questions: how to demarcate lands between the forest department, farmers and gram sabhas; how to protect the claims of tribals and others who had been cultivating the land for ages but without any documents; how to secure land rights in areas which had been suddenly described as reserved forests; how to resolve conflicts among farmers; how to fight the thousands of cases launched against villagers; and how to ensure proper compensation and rehabilitation. With the assistance of the Morcha more than seventeen thousand tribal and other farmers were able to establish their claim over land.

A study and survey of the land issue in Kaimur region by the Morcha brought to light many revealing facts on land alienation in Mirzapur district. Before independence, most of the rural areas in the

United Provinces were governed by the UP Tenancy Act, 1939, but Kaimur region was under the Agra Tenancy Act, 1926. The former gave certain specific rights to the tenants on zamindars' land but not to the latter. Uttar Pradesh state, after independence, saw a partial implementation of the Zamindari Abolition Act, but the areas under the Agra Act were left out for a long time. Even when the land reform acts were implemented, Kaimur region was excluded from their purview for no stated reason.

In October 1953, the state government issued an instruction to the Chief Forest Conservator and the District Magistrate, Mirzapur, to prepare a list of land under the control of the Forest Department. In February 1972, the National Agriculture Commission suggested that unless there was land settlement in the region, agricultural development was not possible. In June 1981, the Revenue Department, UP issued an instruction that alienation of tribal land in southern Mirzapur should be stopped immediately. For this purpose, tribal possession of land had to be registered on the basis of their oral testimony. Land records had to be re-examined and in disputed cases, the tribals' actual possession had to be given priority. These suggestions and instructions were never implemented. After a detailed survey of five villages, the Morcha moved the Supreme Court. The court gave a stay order in July 1983, to the effect that as long as the land survey was not complete in all the villages of Robertsganj and Dudhi tehsils, there would be no displacement of villagers from their possessed land. There would be a time-bound survey and recording of lands and the state government was to constitute a legal tribunal to resolve the long-pending land disputes.

The Maheshwar Prasad Committee, which was constituted by the state government to investigate the matter, submitted its report in December 1983. The committee accepted that there had been no survey and settlement of land in the area, there were serious irregularities in the land records, and that the forest department had extensively violated the rights of the farmers and local people in reclaiming the lands in the forest areas. The committee suggested the constitution of a special agency to do an on-the-spot survey, recording and settlement in two years time. Tribals and farmers were to get free legal help in this venture. For the next three years, nothing happened at the governmental level.

The Morcha, however, kept up its momentum. Meetings were being organized at the village, block, and district levels. Memoranda, delegations, letters, petitions, etc. were part of the campaign.

A massive demonstration at Renukut on 7 May 1986 was a turning point for the movement.

On 5 August 1986, the government announced the constitution of a special agency headed by a Special Secretary, but neither its time-frame nor terms of reference were specified. The villagers then took to agitational methods. From the midnight of 1 October, they blocked all the roads and rail lines in the region. *Chakka jam* and *jail bharo* were the slogans of the Morcha. Hundreds of people were arrested, *lathi*-charged, detained, and implicated in various cases on 2 October, the birth anniversary of Mahatma Gandhi. Dozens were injured in police action.

The Morcha once again moved the Supreme Court pleading for the implementation of the earlier judgement. In its order, given on 20 November 1986, the court said that there was to be a time-bound land survey and settlement in Kaimur region; tribals and others had to be given an opportunity to file their claim; a commission was to be appointed, to which a member of the Morcha was to be nominated, to oversee the survey and settlement work. The state government was ordered to appoint five additional district judges and enough number of survey officers, and to deposit Rs 5 lakh in a free legal aid cell, which would give support to every villager filing his claim.

A unique exercise was now on in the region. Five IAS officers, five deputy collectors, five assistant forest conservators and 800 survey employees were occupied with land survey and recording, to be followed by land settlement. The issue of settlement was, however, still uncertain, because the core principle on which the ownership of land between the different contending parties would be decided was not spelt out.

Tribals and others living in the forest region see land as as important as the forest and want a clear-cut ownership right over it. 'Forests of this region have been ours for ages. Besides, we need the land and its produce for many purposes. The forest is like a life partner and land, our best friend', commented Bechan, a poor tribal, expressing his strong sentiments on the land issue.

Forest Land

Days and Nights of Love and War

In early February 1996, thousands of tribals from remote corners of Udiapur, Dungarpur, Banswara, and Chittor districts laid seige to the Tribal Commissioner's office in Udaipur city. They walked down the main streets of the city, encircled the entry and exit gates of the office and blocked the roads. For three days, they shouted slogans, sang songs, and made speeches during the day, and at night, with some makeshift arrangements, they cooked food and sang songs.

Faced with the threat of eviction from their land and villages by an overzealous government seeking to fence off forests, the tribals were up in arms against the Forest Department and the state government. They were supported by all the major opposition parties in the state. Their demands included regularization of all pre-1980 encroachments on forest lands based on the notifications of the central and state governments, a special time-bound campaign for the identification and regularization of encroachments, recognition for the Jungle Jamin Jan Andolan (People's Movement for Jungle Land), comprising tribal, environmental, social and non-governmental organisations, and citizens, to plan, implement and monitor the campaign, recognition of the *gram sabhas'* opinion for identifying encroachments in the absence of written records and an end to all eviction procedures and harassment by officials until the process of identification and regularization was over.

After three days of the blockade, the Tribal Commissioner's office gave them a written commitment that a survey involving representatives of villages and the andolan would be conducted immediately regarding the pre-1980 encroachments on forest land and the report submitted to the state government, that the opinion of the *gram*

Published in the *Frontline*, 28 June 1996.

sabhas would be taken into account in the regularization of encroach-ments, and that there would be no eviction or displacement of tribals till these processes were completed.

The tribals called off their agitation, but continued with other programmes. The andolan intended to organize a series of village, block, and district level meetings of tribals to make sure that the government fulfilled its commitment. They would return to the city after three months to voice their protests if the survey and regulari-zation was not completed by then.

Rajasthan has witnessed intense conflict between tribals on the one hand and the government, the Forest Department and the local authorities on the other, over rights to resources. A spokesperson of the Union Ministry of Environment and Forests pointed out that the percentage of land under forest in Rajasthan when the state was created was 11.75. Now it is only 9.25. The main reason for this reduction is the release of forest land for various non-forestry pur-poses, such as settlement of displaced people, industries, hydel pro-jects, and regularization of encroachments. H.M. Bhatia, Chief Conservator of Forests, Western Circle, Udaipur explained how the Western Circle extended over 34,920 sq km, of which the forest area was 8084.70 sq km, or 29.5 per cent of the geographical area of the state. But the actual forest cover was over a smaller area. A lot of the forest area had been depleted and some had degraded badly. Per capita forest area in the various forest divisions varied between 0.10 hectare and 0.68 hectare. The forests were burdened with rights and concessions given to the people, such as free collection of grass and dry fuelwood, free supply of timber for houses, buildings, agricul-tural implements and grazing of livestock. Problems like encroach-ment, illegal felling, and mining were further aggravating forest degradation.

Tribals and their organizations, social activists, and environmen-talists supporting their cause presented an altogether different pic-ture. Kishore Sant, an environmentalist and social activist, alleged that the Rajasthan Forest Act, 1953 abolished the rights of tribals in the forests and converted these into concessions. The Forest Depart-ment did not follow the procedures laid down to record and regular-ize the claims of tribal people. As a result, thousands of tribals live in the forests and cultivate lands without any rights. The Forest Conser-vation Act, 1980 has placed restrictions on the use of forest land for non-forest purposes. The land cannot be used without the central government's permission. The basic question, according to him, is how the forest lands were notified and demarcated and whether due consideration was given to the people's rights.

There are several instances of tribals not being aware of forest area notification. Neither were their rights taken into account nor procedures followed to regularize their possession of land. The encroachments on forest land continued unabated. The Union Ministry of Environment and Forest issued a circular (No. 13–1/90 FP (1), 18–9–1990) urging the state to regularize pre-1980 'encroachments'. According to an unofficial estimate, in Udaipur division alone there are more than 32,000 pre-1980 encroachments. A recent andolan survey in Kotra block of Udaipur district suggests that a majority of encroachments took place during 1961–80, most of them extending to between one and ten bighas. Of the 795 encroachments 416 are of 1–5 bighas and 340 of 6–10 bighas. During 1961–70 106 encroachments took place and 515 during 1971–80. Most of the encroachers were landless tribals.

Thirty-five households in Gudha village of Garwa block had been living there since 1971, having occupied forest land. For the past few years, forest officials had been regularly collecting fines from them. Then, in December 1994, all of a sudden, the forest officials began to construct a wall on the land. When the people protested the construction was stopped. But on 31 May 1995, a large number of armed police and forest officials beat them up. Now they were still living in their own houses but unable to farm.

In Nathwara, Govardhan's father had been cultivating three bighas of forest land since 1958–9. On 5 June 1995, the police came and asked Govardhan to show the land records. He did not have any. They destroyed his crop and fenced the land. Since he had no other land he continued to live there, in constant dread of harrasment and eviction. With the inception of various forestry projects supported by international agencies—notably the Aravalli Project (1990–8) supported by the European Community costing ecu 23.30 million, and the Forestry Development Project (1995–2000), backed by the Overseas Economic Cooperation Fund (OECF), Japan, costing Rs 139.18 crore—pressure on tribals and their rights is mounting. There has been much emphasis on fencing or constructing boundary walls around the forest.

By August 1995, only eleven encroachment cases had been identified by the government-appointed committee whereas the process of denotification of forest land for mining leases was completed within four months in 1995. S. Ahmed, Commissioner for the Division as well as for Tribal Area Development, said that they had recently identified 80 cases and submitted details to the state government. The state government's circular said that only landless tribals who had the penalty receipts for encroachments from the Forest Department were eligible for identification and regularization. Sometimes the committee could not meet in the absence of quorum. Though the state government had been regularizing the encroachments from time to time, hard decisions needed to be taken, he said. Regularizing the pre-1980 encroachments might give an impetus to further encroachments.

Representatives of the Andolan met Chief Minister B.S. Shekhawat, Minister in charge of forests and tribal development and others in August 1995 and apprised them of the large-scale evictions. But the harassment and evictions continued unabated.

The issue of forest land rights in Rajasthan is a part of a larger problem confronting many areas in the country. According to Dr D. Saldanha of the Tata Institute of Social Sciences, Mumbai, who has done a detailed study on encroachments in the tribal areas of Thane district in Maharashtra, such encroachments are a symptom of larger socio-economic problems. They are an effect, not a cause, of deforestation. The tribals' symbiotic relationship with their environment has been ruptured and it is a historical process. They have no alternative but to encroach on forest and grazing lands. Regularization of encroachments is not a solution to the basic problems affecting the tribals, but in the context of the alienation of tribal lands outside the forests and in the absence of a concrete plan for the involvement of indigenous communities in the protection and regeneration of forests, regularization of eligible encroachments becomes an inevitable necessity. What today has become 'illegal encroachments' on state forest lands were originally 'legalized encroachments' under the colonial state on the lands cultivated by the tribals within and around their villages. *The Report of the Bombay Forest Commission 1887* (Volume 1, page 31), acknowledged that by a simple notification, nearly 4,01,566 acres (1,60,624.4 hectares) of community lands and free grazing lands were converted into forests. These lands formed nearly 50 per cent of the forest area of the district.

Dr Saldanha, who was a member of the inquiry committee appointed in 1992 under the instructions of the Supreme Court to investigate the claims of tribals on forest land in Thane district, further argued that the absence of relevant documents was no indication of non-eligibility, but was only a result of a chaotic historical process related to the subsistence needs of the tribals. The very fact that these adult encroachers had survived long enough to put forward their claims suggested that their encroachments had subsisted well from their early youth.

The judgement of the Supreme Court in *Pradip D. Prabhu and others vs State of Maharashtra and Others* (Writ Petition No. 1778 of 1986), had set a precedent for tribal organizations all over the country when pleading their individual cases. In this case, filed as a public interest petition by tribal organizations in Maharashtra for legalizing their claim to forest land, the court ordered the governments of Maharashtra and Madhya Pradesh to appoint responsible officers to decide on tribal claims. 'While determining the rights of tribals for regularizsation, they shall be given an opportunity to be heard by the officers concerned and also to adduce evidence in support of their claims. We further direct that till the cases of tribals concerned are finally disposed of, they shall not be dispossessed from their lands which are in their possession', the court ordered.

Forest administrators continued to advocate the 'need to take a hard decision on this question once and for all'. E.P. Thompson in *Whigs and Hunters* has said, 'Claim and counterclaim had been the condition of forest life for centuries ... . Tribals and forest officers had rubbed along together, in a state of running conflict, for many decades and they were to continue to do so for many more... What was at issue was not land use but who used the land: that is power and property right.'

Surplus Land

A Landing after 20 Years

To identify and liberate over 50,000 acres of surplus land in the possession of landlords: this was the new agenda in the Bihar state. As *Bhumi Mukti Andolan* (land liberation movement), it was launched by the left parties, especially the CPI, the CPI(M) and the Indian Peoples' Front in 1992. After almost 20 years, the state leadership of the left parties once again raised the banner of land liberation in recognition of the numerous land struggles going on in the state.

The land issue has over the years taken a complex form. Land liberation movements over surplus, *bhoodani, benami* or government land stretch over a long period of time and have brought up many questions of land management, housing rights, relevance of laws, etc. As the CPI(M) leader Subodh Rai admitted, the task was not completed with the liberation of the land, be it the surplus land of the *zamindar*, the *gair-mozurva* or undistributed ceiling land. More importantly, these land liberation movements faced challenges of sustainability and development. Experience revealed that these movements gave rise to a plethora of events, which needed to be taken up simultaneously.

Kahalgaon subdivision of Bhagalpur district consists of vast diara region where regular floods and land erosion often displace people temporarily or permanently. In the 1980s, after the Kahalgaon NTPC Super Thermal Power Project promised to start a new phase of industrialization, businessmen and traders, in anticipation of rising values of agricultural land, invested huge amounts of money in real estate in the villages surrounding Kahalgaon town.

Some three kilometres from Kahalgaon town is Gangaldai village,

This is a substantially revised version of the paper originally published in *Economic and Political Weekly*, 22 August 1992.

on the Kahalgaon–Vikramshilla road. In March 1990, villagers from the surrounding diara captured around 52 bighas of land in Gangaidai, built their houses and called the entire area Gangaidai Naya Nagar. The land was owned by the Bihar government and one local businessman, but in reality, the government land was also commanded by the businessman. At last count, there were 628 houses in Gangaidai Naya Nagar, most of the households being landless or of small farmers. Bajrangi Das, 70 years, old from Aamma diara narrated how he lost count of the many times he was displaced by land erosion. Being landless they had to live on sharecropping and the wages from farm labour. He and his family had now resolved that even if the police or army assaulted them or fired, they would not move from there.

In Chetharia Peer, on the Bhadeh–Barahat road near Kahalgaon Town, one acre of prime land was owned by a local export business house. The displaced villagers from Bakia diara, Toffir diara, and Jatooya diara 'liberated' this one plot in 1991 and built houses there. In Sukumari *pahar*, displaced villagers from Rani, Khabaspur, and Ekchari diara settled on 2.5 acres of government land earlier grabbed by a local businessman. In Ram Rai Khurhari eighty households from the diara 'liberated' the six bighas of surplus land owned by a landlord. The district unit of the CPI(M)-affiliated Kisan Sabha initiated all these struggles. The Land Liberation Movement has taken quite an organized form, with active party support. A series of demonstrations, rallies, and dharnas have been organized to challenge the might of the landlord lobby.

In Gangaidai, police intervened in support of the landlord and many villagers were arrested. Landlords lodged several cases in the Bhagalpur District Court against 21 villagers. In Chetharia Peer, businessmen again prompted the police to take charge. In 1992, an entire new *basti* was burnt to ashes by conspiring businessmen. In Sukumari Pahar, local henchmen and anti-social elements were trying to terrorize villagers.

A big trading concern in Ghogha Bazar owned nearly two hundred acres of land near Khiridand, a typical backward village located about 40 kilometres from Bhagalpur. The sharecroppers of the village had been working on this land for nearly twenty years. Some years ago, the landowning concern started selling its land in bits land pieces, which resulted in nearly thirty sharecroppers being evicted. The sharecroppers then organized themselves under the banner of the Kisan Sabba to protest against the eviction and to press for recording their sharecropping rights. A number of meetings were organized in the village. Several villagers filed legal cases to protect their rights over nearly sixty acres of land, but the administration did not take up their complaints and demands. The villagers then decided to claim the remaining land, thus saving it from further sale, and also decided to claim the crop standing on the land. One morning in December 1987, armed police came to arrest their leader Srinivas Mandal. The village women surrounded Mandal to prevent his arrest, getting hit with rifle butts by the police. The police left, but returned two hours later with reinforcements. In the ensuing police firing, three women were killed and two sustained bullet injuries. Several other men and women were also injured. The police however lost the battle once again. The police and the trading families then began to implicate the villagers in false cases. Terror tactics were adopted. The villagers described how during these years some part of the land also went out of their hands. Sharecropping' rights were never registered and every year the distribution of crops became contentious. The legal process was long drawn and as a result the villagers only succeeded in getting hold of a small area of land, while the larger part of it was still beyond their reach.

In Nayan Sukh Nagar Musahari village, 15 kilometres north-east of Kahalgaon, all the fifty families are musahar by caste, most of them landless and without any other asset. One retired police officer from Bhagalpur owned a vast area of land near the village, which Musahari residents cultivated as sharecroppers for twenty-five years. The landlord, with the connivance of the *patwari* and local police, also grabbed six bighas of land of the musahars in the village, and the oppressed villagers worked as sharecroppers on their own land. After the Kisan Sabha came on the scene in 1990, the villagers decided to liberate their land and stopped giving the cropshare to the landlord. The landlord evicted the sharecroppers and barricaded the land. The villagers brought down the barricade. Then the police intervened. In May 1991, an armed group of the landlord came to till the land. The resisting villagers were fired upon, seriously injuring one person. Many women were severely beaten up. The police arrested eleven persons. Section 144 was promulgated in the area and the CRPF was summoned to march through the village. Arrest warrants were issued against the Kisan Sabha leaders and for months the villagers were terrorized. Now the six bighas of liberated land are under police control, the villagers are tilling the land and sharing the crop.

In Dewari, a tribal village of seventy houses, a landlord captured about fifty acres of land of the villagers a long time ago. When the

Kisan Sabha organized the tribals in 1991, the whole tribal *basti*, armed with their traditional bows and arrows, took charge of their land and the standing crop. The tribals successfully foiled attempts by the police to intervene in favour of the landlord. They also intended to initiate the struggle on other lands as well, like gair-mozurva, common land, etc.

The story of land liberation movements has instances of both victory and defeat. But there is a common thread. If an organization takes up this issue and struggles continually, the villagers acquire great strength and as a rule become assertive. The movement has moved beyond mere slogans or symbols. Given the social tensions it has generated in the state and its spirit of militancy, it is important to continuously follow them up and understand their complexities and varied forms.

❦

No Way to Employment

'We are two in one—Musahars and *mitti* (soil). Digging soil is like a physical exercise for us, much better than sitting or lying. When we don't do it, we feel tired,' Asharfi Sadai says, sitting idle in his village Sirpur Mushhari, six kilometres east of Jhanjharpur town in north Bihar.

Musahars have dug deeper and faster, and carried on their skills of measuring and assessing the quality of soil for generations. Asharfi Sadai represents many thousands in the area who took pride in their traditional skills. He did not need an inch-tape, any modern measuring device or the advice of an engineer. He did not remember the large numbers of dams, ponds, bridges or multi-storeyed buildings where his skills and labour were applied in assessing, digging, or measuring the soil.

Now he gets less work linked with soil. Like many others he had no land. Most of his life having passed working with soil, his choices are very limited. However, being enterprising, and unafraid of learning new things, Asharfi has now started weaving baskets though the going is tough. He does not own the bamboo trees. They belong to somebody else. The basket market is uncertain. If somebody provides the bamboo, he weaves it for them. Otherwise, Asharfi searches for soil work. He has been to cities, like Darbangha, Samastipur, Muzaffarpur and sometimes even to Patna. In spite of various on-going construction activities, there is less and less demand for labour like his. 'I see how for an activity like assessment and measurement, we are no longer required. Further, the digging work in big project works is being done by the machines,' he explains.

There is less work, and less wages, in the available construction jobs. The wages for house construction activities in towns are Rs 30 to 40, which is less than the prescribed government wages.

This is a substantially revised version of the paper originally published in *Labour File*, January 1998.

Musahars are a Scheduled Caste whose total population in Bihar, according to the 1981 census, is 13,91,000. They are widely distributed in several districts—Madhubani, Muzaffarpur, Darbhanga, Champaran, Hazaribagh, Santhal Pargana, Bhagalpur, Munger, Purnea, Gaya, etc.. In 1891, Sir Herbert Risley wrote, 'Musahar are an offshoot of the Bhuiya tribe of Chota Nagpur.' Deriving meaning from the word Musahar, some say they signify flesh seeker or hunter (*masu* meaning flesh, *hera* meaning seeker), while others interpreted their name as rat-taker or rat-eater (*musa* meaning rat).

Musahars' traditional job market has been squeezed in some other not-so-obvious ways, especially in the recent years. At the village-level developmental works, under the *Jawahar Rozgar Yojana* or other schemes, there are a number of works related to digging and carrying of soil, which were being done by Musahars before, but not anymore.

Harhar Sadai, of the same village said how now they only do the digging. The loading, unloading and carrying of the soil—are all being done by trucks and tractors, contractors and their labourers. Here the commission rules the game. Surya Narayan Sadai of village Bheja in Madhapur Block complained of similar incidents. Trolleys and contractors have taken over the scenario, leaving very little scope for them.

Musahars have always lived here without land and cattle, but they have survived as agricultural labourers and sharecroppers. Even women have been participants in this work. Bhola remembers how initially in the 1970s and 1980s, various family members were working together here. They were also saving. More labourers were required. This is no longer possible.

A worsening ecological situation, with floods, shifting course of rivers, adverse fallout of drainage congestion, embankment, and water logging, has had serious impact on the rural agricultural scene of the region since the early 1970s. Sonai Sada, a villager in Bakunia Engineer village of Saharsa district, has no hesitation in referring to embankments as 'the dreaded killer of our agriculture and land.'

In the more recent past, the dispossession of Musahars from rural land and employment has taken a new turn. In recent years in Bakunia, landlords have purchased three tractors, and hire tractors from outside as well. One landlord rents it to another for ploughing, at the rate of Rs 150–170 per bigha. 'It has become easier for landlords, as now they do not require labour, cattle or plough for a major part of their farming,' says Tithar Sadai.

Hence now there is work available locally only for three–four months and primarily for womenfolk. Bindo Sadai, another Musahar labourer in the village, confirmed how sowing and harvesting, and in between, some other works related to standing crops, were barely enough for the womenfolk. Musahar men were facing more and more unemployment. The new landlords, the Yadavs, did not want them anymore on a long-term basis.

Musahar men and women work from six in the morning to one in the afternoon. They get a meal of two *rotis* at 8–9 a.m., and their daily wages were 2–3 kilogram of maize or wheat, which cost approximately Rs 12–13.

Earlier, there were a variety of other works available in the village vicinity—making of paper packets, knitting ropes, weaving baskets, stitching leaf plates, husking rice and preparing eatables from it. But these works have virtually died out today. 'Rice mills have opened both in the village and at the block. Their owners are not Musahars or other poor. So landlords prefer to go to these mills, rather than calling Musahars to their houses, or giving gears to them for manual processing,' informed Surya Narayan Sadai in Bheja village.

There are a number of individual responses in the region among the Musahars. Some are becoming rickshaw pullers in greater numbers, in Jhanjharpur, Madhubani and other places. Others have turned to vending here and there, and sell on piece rate basis whatever they can get from shop-owners and retailers. In Kheri village of Lakhnaur block, Mahakant has opened a cycle-rickshaw repair shop since 1994, 'I was a sharecropper before. But that was not working.' Similarly, in Bakunia Bichali village, Pulakit Mandal is a tailor today, with a sewing machine of his own. First he migrated to Punjab in 1988–9; sustenance there proved really difficult. Somehow he managed to purchase a sewing machine in 1993 and since then, this has been his occupation. Normally he gets work for 10–12 days a month and is able to earn Rs 800–900. Some others have also started the tailoring work. It is not easy to get work and earn cash. But he does not want to go to Punjab anymore.

Musahars as such never owned soil, since land never belonged to them. However, their talent of working with the soil was their very own. And this relationship was an integral part of the use of land here. This phase now witnessed is not just one of replacement of that skill by new equipment. But that the control of this equipment is no longer with those who used to work on land with the skill acquired for generations. It is run by trolleys, tractors or construction companies. This would also open up new hazards for land here in future.

Remapping Ralegan Siddhi

At village Ralegan Siddhi, district Ahmadnagar, Maharashtra, it is still a couple of hours to dawn. At Sri Sant Niloba Rai Vidyalaya, the hostel boys are already geared up for the day's work. They are dressed in khaki half-pants and white shirts. Some wear Gandhi caps. They pass in single file, on their way to work in the school's fields. Some clean the schoolrooms. Outside the school premises some women are carrying baskets and a few men are cleaning the streets. Yet others are moving towards the village land. From the public address system installed in the school, invigorating popular *bhajans* and Hindi songs are booming: 'Vande Matram, Vande Matram!', 'Sabarmati ke sant tune kar diya kamal.'

Ralegan Siddhi and Anna Hazare are household names in Maharashtra. Their land and water management, experiments in rural development, self-governance, everyday beliefs, practices and cultural ambience and anti-corruption campaigns have often caught the nation's attention.

The village of just two thousand people boasts tremendous economic development owing to its capacity to deal with environmental issues collectively. It professes to combine materialism with spiritualism, religion with patriotism, and environment with development.

A village where Anna's words are the rule, he has an acknowledged right to command. The villagers show an acknowledged obligation of obedience.

✝

In a district that is hilly, dry, and dusty, Ralegan Siddhi stays fresh and green. A few years ago, hardly any trees had been left in and around the village. Even the 400 millimetre average annual rainfall led to further land erosion. During the summer months tankers provided drinking water. There were a few wells which could irrigate a maximum of 24 hectares of agricultural land, according to the 1981 census. At regular intervals, especially during a drought, many villagers migrated for wage labour. Ralegan Siddhi was, by all accounts, a village living on the edge.

The drought of the early 1970s affected the entire region acutely, and though some government relief programmes came to the village, they were poorly conceived and badly implemented. A percolation tank was planned for the village, when it was finally constructed in 1983, did not work, because the water seeped away from the bunds.

In this village of 285 households, except for thirty scheduled caste households, all are Marathas. Of a total cultivable area of 651.13 hectares, only 30 households have more than four hectares, 62 have between two and four hectares, and the rest have either one or two hectares or no land at all.

Under the leadership of Anna Hazare, the villagers evolved and implemented a different path of watershed development. They constructed storage ponds/reservoirs and nala bunds in a series, along the 30–45 metre high hills surrounding the village. The nala bunds were of different kinds, big and small, open and underground. Between all these, extensive plantation was undertaken. Soon the hills and the nearby areas were all covered with bunds, trenches, nalas and plants. Thirty-one nala bunds, with a storage capacity estimated at 2,82,182 cubic metres and covering an area of 605 hectares were completed. Four lakh trees have been planted till date. Significantly, the bulk of this has been accomplished utilizing the financial resources coming to the village under the government's soil conservation and social forestry programmes.

'Water should not be seen on the surface, it should be trapped and kept below', was the guiding principle in Anna's words. This type of water conservation led to many positive developments in the course of just a few years. The extent of arable land as well as the yield increased, and the groundwater level, which was 100 feet below, came up to 40–50 feet. The wells and ponds also got filled and now even in a year without rain, the village is not without water.

Anna has persuaded the villagers to experiment with water and land in multiple ways. Fifty-three small farmers with adjacent lands dug and owned eleven wells jointly, to irrigate more than a hundred acres of their land. In another instance, 103 farmers were convinced to form a cooperative, to pool their resources and produce, to cut the cost of the inputs. Further, the defunct percolation tanks in the village were repaired through voluntary village labour. Above all, the distribution of water among the villagers has been regulated. Unless all farmers get their first round of irrigation, nobody can access a second round of water. Landless families are compensated in cash or kind by those who use irrigation water. Those who dig individual wells share its water with neighbours.

Anna Hazare's slogan of *shramashakti dwara grameen vikas* (rural development through labour) has been adopted by the state government to design a rural development programme for the entire state. In this programme, at least two persons, in the age group of 12–58, of each family in the village have to do *shramdan* for a day once in a month or alternatively contribute a day's salary to the project fund. There is a guarantee of attendance of at least fifty labourers from the village in which work has commenced. Developmental work has to include forest farming, digging of wells and construction of percolation tanks or bunds. To adopt any programme at least a two-thirds majority of the gram sabha is required, together with a ban on felling of trees and free grazing of cattle in and around the village.

The stress in all these self-help efforts is on the concept of *dan*, i.e. offering, initiated by the villagers themselves. The school buildings have come about largely through such efforts. The school has about 450 children, 135 of whom live in the hostel. The school caters for studies up to the tenth standard. The villagers are organizing another round of dan, to upgrade it to a higher secondary school. Dropouts and rejected children get preference in admission.

Shramdan is part of the daily school curriculum. Every morning children clean up some portion of the village for half an hour, clean the school latrines, and once every week clean public toilets of the village. During the rains, they engage in plantation work for days together.

Dan is also the basis of the village's 'grain bank'. When the village improved its agricultural output, most of the farmer households, under Anna's instructions started contributing a share of their harvest regularly, and a 'grain bank' came into existence. This bank is now full, and it even sold 100 bags of grain in the open market in 1989–90. The earnings from this went to the school. The main purpose of the bank is to give minimum food security to needy families in the drought-prone and dry region. Those in need can take a grain loan from the bank, and in the subsequent year return the loan with 10 per cent more grain.

Anna Hazare encouraged people's active participation in planning and decision-making of various programmes, so that 'the village is built through the creative, productive and innovative hard work of people themselves'. The village has fourteen *vividh karyakari* societies, dealing with forestry, water, the co-operative, the school, etc. All nine members of the village panchayat are women. They were elected unopposed according to the wishes of Anna, as he wanted women to contribute significantly to village development. Among other things, they have ensured the payment of equal wages to women labourers and raised the consciousness regarding the education of the girl child.

Partly, the success of Anna Hazare's development model is also attributable to a high level of adherence to many 'dos' and 'don'ts' in both public and private life, enforced with punishment. In the 1970s when Anna Hazare started his endeavour in the village, he came down with a heavy hand on liquor *bhattas* and alcoholics, to the extent of personally tying a person to a pillar and flogging him with his army belt. No shop in Ralegan Siddhi can sell bidis or cigarettes. Film songs and movies are not allowed. No sugarcane cultivation is permissible, since it consumes a lot of water and also gives ample leisure to the farmers. *Kurarbandi* (ban on tree felling), *charibandi* (ban on tree grazing) and Shramdan became integral mottos of the programme.

The village, which had a per capita income of Rs 252 in 1976, now claims that it has risen to Rs 2000. H.V. Huprikar, branch manager, Bank of Maharashtra, in the village, said: 'The village has some money now and the people are conscious of savings. The total savings of the bank branch are Rs 42 lakhs, of which Rs 23 lakhs is that of the village. Even school students have accounts under the "minty" scheme and have deposits of nearly Rs 18,000. The recovery of crop loans in the village is about 80 per cent.'

The villagers have shown competence in managing a series of projects, artefacts and technological measures. They have grafted the drip irrigation system, solar panels and *gobar* gas plants. Ninety-three acres of land are being covered with drip irrigation. The village has two biogas and thirty-two *gobar* gas plants. Streetlights with solar fittings are being managed by the locals, supplemented, if necessary, by service-providers from outside the village.

+

In the early 1990s, Ralegan Siddhi was famous not for its land and water development programme, but for a massive campaign by Anna Hazare against corruption in rural development programmes. Contracts for developmental works were awarded by government departments and schemes were officially completed, but the actual work was rarely visible. Anna Hazare decided to take up the issue of grassroot corruption; he began his groundwork by writing letters to all the gram panchayats in the district seeking information on two things: whether the rural water supply scheme was working in their area, and what had been purchased under the social forestry schemes and at what rates. It emerged that the rural water scheme was not working in 165 gram panchayats and in social forestry, many useless instruments were being purchased, at highly inflated rates.

When Anna Hazare approached higher government officials with these facts, the Chief Secretary of the Government of Maharashtra gave an assurance that action would be taken, but nothing came of it. As a lone crusader, Anna began his silent fast in December 1990, and returned the famous Indira Priyadarshni Vrikhsamitra Award given to him by the Government of India. He declared his intention to also return the Padmashree Award by 26 January. Then came the proposal of the state government to constitute inquiry committees. Anna Hazare broke his fast on 17 January 1991 at a meeting attended by more than six hundred representatives of social-political organizations from all over the state.

The inquiry committees found many of Anna's charges to be legitimate, and initiated actions against some government officials concerned. In Mumbai, A.B. Gokak, Secretary, Forest Department, Government of Maharashtra, admitted in February 1991 that allegations of corruption in the social forestry schemes of Nasik and Kolhapur division were found valid in the preliminary inquiry. The purchases were below standard or of no use, their rates were high, and various rules were violated. Two officers were charge-sheeted. It was also decided to extend the inquiry to the entire state. Anna Hazare had complained of many irregularities in water supply schemes also. On inquiry, it was found that 30–35 percent schemes in the villages were non-functional. The committee constituted to inquire into the water supply scheme also observed that all the water schemes were based on wrong assumptions about the need of 40–litre water per capita per day and a regular supply of electricity for sixteen hours. In actuality, a person needed 70 litres of water whereas electric supply beyond 8 hours was unnecessary.

Anna Hazare now put forward a demand that is yet to be accepted by the government. He asked for an Inspection Committee of five members, not belonging to political parties, to be constituted at the level of the village, block and district, which has government approval to inform the public about government schemes, inspect their implementation, and monitor the tenders, contracts, and other things. If the government did not heed this vital demand, Anna said he would tour the entire state, form committees on his own to monitor the projects, and inform district magistrates about the corruption and other malpractices. If no action was taken, he and his supporters would launch a *morcha* and *gherao*. They would even take the law in their own hands and beat up the corrupt officials in the open.

+

A middle-aged bachelor clad in *khadi*, Anna Hazare lived in the village *mandir*, which he renovated out of his savings, after retiring from the army. His power over the villagers seems all encompassing and came through a long and dynamic process of interaction with his village, activists, bureaucracy, and the government. This process was characterized by projections and counter-projections of religion, culture and tradition; and by deployment of natural and human resources, including persuasion, coercion, and possibly suppression. His personal qualities have been exercised predominantly in traditional ways like rebuilding the temple, owning no private property for himself, and having no distinction between the private and the public. His authority is based on a traditional structure of power, such as respect to the elderly, patron–client relations, and discipline in social norms and behaviour, especially at the local level.

At Ralegan Siddhi, he adopted two routes simultaneously: imposition and transformation. Imposition, because 'Rural India is a harsh society. If you want to change things, it is sometimes necessary to be tough. It cannot be done by Ahimsa alone'. Transformation, because 'Spirituality is not sufficient. Roti is important for an empty stomach. When many people slept on empty stomachs in this village, nobody was willing to follow us. But after the improvement in their economic situation, they got ready for shramdan and other things.'

Since the very beginning, Anna Hazare has been successful in defining moments and issues which can generate deep emotion, and galvanize enormous support for a cause. This moment may be a single event, or a string of related events. But they bear enormous moral authority that legitimizes not only the cause but also its initiator, while delegitimizing others.

Some years ago, faulty transmission and low voltage caused several pumpsets to burn in the village. In spite of repeated requests by the villagers to install a substation, government agencies took no action. Anna Hazare went on a hunger strike, and had to be hospitalized after nine days. This led to a *rasta roko andolan* by the villagers. In the ensuing police firing four persons were killed. Later the government sanctioned the money for setting up the substation. On another occasion, the Education Department refused recognition to the school which the villagers had constructed themselves. It was alleged that the Education Minister was penalizing the villagers for not voting for him in the election. After several futile meetings, Anna Hazare threatened to go on an indefinite hunger strike. 'If I die, do not cremate me, but bury me under the steps of the village temple', he told the panchayat. The deeply moved and surcharged villagers gheraoed the zila parishad the next day and obtained recognition for the school.

The anti-corruption movement is a socially and politically defining moment for Anna Hazare. Inspired by a continuous and common set of grievances in his district, as well as in others, the movement extended the reach of its belief enormously, and contested others sharply. 'Power and politics cause corruption. Those who wish to involve themselves in our anti-corruption movement, will have to take pledge not to get involved in party politics, nor to contest elections', explained Hazare.

Anna has a number of resources at his command. Though issues related with land and water were given maximum expression, he also dealt with many others like education for children, increasing the participation of women, anti-liquor campaigns, banking for the poor, to name a few. Thus his appeal extended to different audiences at separate moments, and enhanced his following and authority.

Anna prescribed certain values and projected goals for society: 'People should have good *samskar* to do service. They should believe in *nishkam karmayog*'; 'It is necessary for people to retain their humanity along with prosperity'; 'Prosperity and the sharing of wealth brings equity'; 'Differences between the rich and the poor will remain, but today the poor man has grain in his house and he can sit with the *Patel*, because he no longer has to beg before him. With *jan jagriti*, differences will go.' These are the essential moral values or ideology of Anna which specify how things ought to be and work as a normative regulation for society. Through a central value system, they exerted a powerful impact on the determination of goals, and the means to achieve them. Its institutionalization in a village created a social cohesion on the surface.

As Anna is deeply committed to his values and beliefs, he earnestly wants to pursue their replication. He now proposes to start a rural development-training centre at Ralegan Siddhi. He would like to spend more time visiting other villages. His legitimacy is a complex feature today, in search of constant cultivation. Since he has chosen to work and rework continuously, his life and work will be subject to continuous redefinitions as well.

Inland Fissures

Indigenous purification process

Plankton used for photosynthesis and fish food

Ravines
A Law and Order Problem?

Locally, they are called *'bechiragi'* to mean lightless, lampless. Not so long ago, they were abuzz with life and normal activities. But now these hundreds of villages all over Bhind and Murana districts of Madhya Pradesh lie devastated by the increasing ravines, with ruined houses, ditches, and wasteland.

In Murana district, village Mrigpura is in the process of being swallowed by the ravines. Village Chursalai is only 17 kilometres away from Amba, but has to be covered by foot or camel. Most of it has been destroyed by expanding ravines. The fifty households in the village have no agricultural land now. The villagers either did some cultivation on the river bank or cut forest trees to sell in the city markets. The single well in the village provides water for only about an hour everyday. Villagers have to go to the nearby Chambal river and get water on camelback.

Village Rudawali, with a population of about five thousand and not more than 50 kilometres from Murana is being segmented into three by the ravines. Almost half of the village consisting of houses, schools, and streets are ravines. In the middle of the village there are 50–60 foot deep ravines. Small and big nalas and pits are increasing, facilitating the onrush of water and the formation of more ravines. The groundwater level had gone down to 200 feet. Three men with the help of a buffalo are required to pull water from the wells. In Porsha block of Murana district, Ratanbasai village is now segmented into eight new parts. The ravines have destroyed the streets and roads and it took a tough walk across 3 kilometres to cover the village. Poverty, unemployment, and insecurity apart, the villagers are concerned about their dying social and cultural life. It is becoming a problem to get their children married because the people of other villages do not wish to establish relationships here. Even among themselves, they have difficulty in keeping contact.

Over the years, thousands of hectares of fertile land along the banks of Yamuna, Chambal, Mahi and their tributaries have been ruined by ravine formation in the states of Uttar Pradesh, Madhya Pradesh, Rajasthan and Gujarat. Of late, the process has accelerated more so in the Chambal region, which is the worst affected. According to K.L. Jain, Commissioner, Chambal Division who resides in Gwalior, 948 out of the 2141 villages in Bhind and Murana districts have been affected by ravine formation. The affected area is about 3.107 lakh hectares out of a total area of 16.14 lakh hectares in the Division.

A study done by the Centre for Science and Environment, Delhi reports: 'In the last 30 years, ravines have increased by 36% in Bhind and Murana Districts. Between 1943–1950, every year 800 hectare area became ravines in these Districts. During 1950–75, ravine formation increased, at the rate of 5000 hectare area every year. Its main reason was large-scale deforestation. Only 20.58% land has forest cover in these Districts.' Dr K.S. Senger of Government Girls' College, Murana, estimates that by 2050, an additional 52,000 hectare agricultural land in this region would turn into ravines. Around 1500 villages would be ravine-affected and the per capita land availability would come down from the present 0.33 hectare to a mere 0.124 hectare.

Dr Ram Prasad, Director, Indian Forest Research Institute, Jabalpur explained: 'Less vegetation level on land makes room for the rainwater to flow (or to sweep away) the upper portion of that land. Thus water flows fast, and creates nalas, big cracks and fissures. These develop slowly or quickly depending on the coming of the rainwater, and become ravines. Once even a small ravine is formed, every rain makes it bigger by creating holes in the front as well as the corners. In the specific context of Bhind and Murana districts, the light alluvial soil, coupled with deforestation, increasing pressure of a population unmindful of the flow of water, faulty irrigation projects, and short-term developmental schemes seemed to be the prime reason for the cancer-like formation of ravines.'

In 1970–1, 'ravine agriculture scheme' was launched by the government for restoring shallow ravines for cultivation. Till 1990, only 1175 hectare land had been restored, at an expenditure was more than one crore rupees. A big central scheme also came into existence in 1971, with an estimated expenditure of Rs 1224 crore over a span of 28 years. Each of the four stages of the plan had a component of afforestation, agriculture, pasture, contour building, drainage and small dams. In 1972, the scheme, conceived under the Ministry of Home Affairs was terminated in its first phase itself, after hundreds of dacoits of the Chambal region surrendered at the initiative of Jayaprakash Narayan. Perhaps the government thought its task was over in the ravine region. Apparently for the governments, ravines were merely a law-and-order problem since they were infested with dacoits.

In 1980, with World Bank aid, the government started a project called aerial seed spreading. The aim was to make green every year 12,000 hectares of ravines by sprinkling seeds from aeroplanes. The seeds of acacia were supposedly supplied by a multinational company 'Wimco'. They were of an alien quality, and the project was doomed to fail. Only some patches of acacia plants are now visible. The seeds dwindled away, but the project continued. The Chambal canal irrigation promised a green revolution and increased agricultural production for the region, but the benefits remained limited mostly to the non-ravine regions. Some of the ravine-affected villages now had the additional problem of water logging, because of canal irrigation. Barwai village lost about a hundred acres of land to waterlogging. In Goshpur village, the canal system made the groundwater level so high that water flowed out from the village wells and made the land marshy. At one time, the government officers assured them that if they went in for sterilization, their new problems like water logging and drainage would be solved. Some did go for it, but nothing came out in the end.

At some places, the villagers' self-help initiatives have worked, and at some places amidst miles and miles of dry, lifeless ravines, patches of green, cultivated land stood out. Sirsani village in Morena district turned 15–20 foot deep ravines into plains and started cultivation though the area was under the government. The government gave a reward of Rs 250 per bigha, but it was poor incentive for the hard and risky work of ravine reclamation. The village has more or less stopped efforts in this direction.

In Rudawali, the villagers made a makeshift dam to halt to the ravines from affecting the village, but this was washed away in rains and flood. Nevertheless, the villagers were successful in creating obstacles in several small nalas and stopping water flow within the village. The villagers of Naya Baans did the same. In Mahuwa village, the worried villagers collectively built a boundary wall around the village, thus blocking the entry of ravines. In Palukapura, though ravines had spread around the village, the villagers were successful in saving their houses for many years by making rough dams and stairs within the pits and nalas. Kadaura village of Bhind district, in a sustained effort against the ravines, filled up the village and its border with massive plantation of grasses of several local varieties, which bind up the soil.

The demand to lease out ravine land to farmers was echoed in many villages. In Amba block, some residents of Bilpur and Kuthiana village claimed that some twenty years ago, they had got government lands in ravines on lease. They have made that land cultivable and since then possess it.

Bhai Mahavir, chairperson of the Chambal Valley Peace Mission, closely involved in the ravine region of Madhya Pradesh for more than five decades, expressed the view that the success or failure of any ravine-related programme depended on how far the villagers were involved in it. The only way to ensure their participation was for the government to leave its ownership rights over the ravines and lease them to the farmers. In addition, farmers had to get all-out support.

Bhai Mahavir warned that though the government promises that once the problem of dacoits is over, it would make all efforts for the development of the region, the ravines themselves had become a terror now, and without any genuine developmental efforts, the villagers would face the dual terror of dacoits and ravines in the future.

The Science Centre which worked to popularize science among the common people, organized a national convention at Gwalior. The occasion marked a long-awaited congregation of government officials, political leaders, social activists, scientists, students, local people and a large number of villagers. The Convention suggested many things like construction of small dams to check erosion, plantation of trees and especially grasses of local varieties, lease of ravine lands to farmers, and so on. The Centre had already started 'Stop Ravines Committees' in the villagers, in the ravine-affected areas. Villagers were coming forward and enthusiastically taking part in the meetings organized to discuss the problem and its resolution. 'It seems the signs of life have not disappeared from here. Between the heaps of stones, fissured walls and the dying wells, there still remains a lust for life. We hope this air would once again be heavy with the smell of moist soil, and its evenings be laced with the splintered, rich songs of *Alha* and *Udal*,' dreamt Arun Bhargava, excited by his recent experiences of the campaign.

Today's Halma

According to *halma*, an age-old agricultural tradition in the tribal villages of Jhabua district, Madhya Pradesh, most of the families collected their agricultural instruments and worked jointly on one person's land on a given day. The process was repeated everyday to cover every family. The instruments were privately owned, but their usage was collective. The land was jointly cultivated, but its produce was private. This ensured a substantial reduction in the cost of means of production.

Today's halma and *pian-no-halma* are recent incarnations of this tradition, begun by the Integrated Rural Development Programme (IRDP), and other poverty alleviation programmes in the region. In Gainda village, hundred-odd families joined hands to form a group with support from IRDP and the bank. They jointly constructed dams and *nalas*, so that benefits from the irrigation systems could reach everyone on a long-term basis.

Villagers, as a working group, decided the location and timing of a particular work. They jointly collected the wages and distributed it equally amongst themselves. They ensured the quality of construction and its maintenance. A villager clarified that it was not a co-operative though the bank and government financial bodies wanted the villagers to form one. But they decided against it, as they felt that this could lead to corruption, bureaucratic interference and other problems. 'It would have spelt death for the halma tradition. We are simply a halma group. The administration and the bank recognizes us as such.'

Pian-no-halma is another name for lift irrigation, which became popular among the poor villagers through the poverty alleviation programmes. Under this scheme families owning small plots of land

This is a substantially revised version of the collection of articles was originally published in the *Navbharat Times*, 17–20 June 1990.

come together, plan, implement and reap the benefits of the irrigation facility jointly.

Jhabua is one of the most densely populated tribal areas of the country, where 80.94 per cent population consists of Scheduled Tribes. They primarily depend on agriculture and wage labour; land holdings are small; and land distribution more or less equal. The District Agricultural Officer stated that the district received an average of 30–5 inches of rain annually. Being a stony and sandy terrain with less forests and more slopes, the land does not hold rain water—instead, rain water leads to land erosion. Drinking water and irrigation were serious problems here. For this reason, water here spells power, often controlled by a few. During the agricultural season, the poor depend on the rich for water, and they have to pay for it in monetary and other terms.

As usual, the government departments are confronted with problems they hardly know how to deal with. In this hard, difficult region which suffers from a shortage of water, scanty pastures and poor soil, the officials distribute cattle; where there are several schemes for sustainable development like watershed management and forestry, they only construct roads! However, as the young and enthusiastic District Magistrate, Sanjay Joshi, informs, halma and pian-no-halma have taken root in two villages of the district. In the last six months, 84 schemes have been started under this. More and more villages were now coming forward to embrace it.

Irrigation schemes in the country are often accused of benefiting the landlords, the rich, and the urban population alone. This could also be true for smaller irrigation projects, if they are high cost. Sanjay Joshi said that the cost of a small irrigation project in these villages used to be Rs 30,000 per acre—way beyond the means of the poor tribals and they could only afford it through loans. However, the collective character of halma brought it down substantially to Rs 1200–2000. The government and the bank gave 75 per cent and 25 per cent loans respectively. A poor farmer could now afford irrigation of two crops a year, at a cost of a few hundreds per acre. Water availability also made power-sharing among the poor possible. Not only did they not have to buy water, sometimes the rich had to come to them. This had never happened before, pointed out Haria, the villager.

Halma reveals the myriad possibilities of organizing the rural poor on issues of environmental regeneration, and giving them rights over natural resources. A team of researchers from the National

Labour Institute, Delhi, led by H. Pais and C.S.K. Singh, came to the area in the late 1980s to assess the various poverty alleviation programmes. From the five blocks of the district, they selected two villages each and on the basis of a survey of a total of 340 people, came out with several uncomfortable facts.

The selection of poor families by the district administration was such that many poor families were left out, and many non-poor included; the programme was merely reduced to the distribution of cattle and construction of the roads. In other programmes, like the Rural Landless Employment Guarantee Programme, out of the 19 beneficiaries, nobody was landless and some even had big landholdings, and all they got was an employment for 3 to 15 days, whereas the target was for 100 days. The lead researcher C.S K. Singh summed up the report by pointing out that not only were the poverty alleviation programmes not being able to meet their target of uplifting the poor above the poverty line, in reality the poor were actually getting more destitute and indebted.

While making the report, the villagers were made aware of its various findings, and the researchers and the villagers met at various levels. Group discussions, meetings, workshops and recommendations led to the formation of an organization of rural farmers and agricultural labourers called the *Gramin Kisan Khetihar Majdoor Sangh* in 1990. This non-party, non-funded union of the rural poor first came into being in the villages of Thandla block, and later spread to other blocks as well.

Harbans, an activist of the Sangh, in Thandla explained that many problems, besides the deficient implementation of government programmes were prioritized. The issue of land records, land alienation, unregistered–unanimous land transactions, role of the forest department, irrigation, drinking and dowry, were a few among them. The organization went ahead with whatever it could do on its own, to solve some of these. Alcoholism was found to be a major problem in most of the villages. Thus the organization decided to impose a total prohibition and in case of its violation, a penalty of Rs 40 was usually implemented.

The other major problem was of dowry. The demand for dowry had tremendously increased in the past years, and the villages were taking recourse to moneylenders, sale of land, and forced labour to meet it. The organization fixed a maximum limit of Rs 525 as dowry which the people accepted. The villagers began to address the district administration on wider issues of restructuring of the poverty alleviation programmes, or the implications of irrigation projects. According to C.S.K. Singh, who was personally involved in many of these organized activities, they had first tried to communicate with the administration and also offered to organize camps of villagers and government officials together, to thrash out the problems of corruption, programme implementation and people's participation. It worked to some extent. Sometimes, the organization had to resort to demonstrations, *dharnas* and blockades.

It was now ensured in the poverty alleviation programmes that the beneficiary families were selected by the administration and the organization together. Also that a committee of government officials and beneficiary families jointly did the purchasing of cattle. But many more demands like these remained unfulfilled. Many villages of Jhabua District are now busy exploring and experimenting with new ideas. Halma is taking many shapes. In Jamani village, 60 kilometres from Jhabua 35 tribal families owned 5 to 10 acres of mostly barren, waste land. These wastelands now have a new look. Within the land, there are small soil embankments to prevent erosion by rainwater. And there are more embankments at one end of the land, so that the remaining flow of rainwater was caught there. There is a slightly bigger embankment outside the land area, to store water even there. Whatever water is left is caught in nalas through a stopdam. There are also wells and small ponds.

Of course, the district administration supported this programme, but it was the villagers who went for it zealously. Sanjay Joshi explained to us that this is done to use every drop of water possible and ensure irrigation all round the year. It was an effort to combine land and water conservation, as land itself was treated as a water catchment area. This is known as the 'Jamani experiment', within the halma tradition. In this region, villagers realized that to claim rights over forest and land was not enough, and the administration recognized that a mere expansion in the irrigated land area, rather than solving the land erosion and water scarcity problems, aggravated soil degradation. How long this will sustain remains uncertain.

Wetland
A 'Him or Us' Situation

The wetlands of Calcutta are shrinking amidst serious controversy. The 'East of Calcutta Wetlands' movement, which took up the issue, poses a serious resistance to the real estate business in this metropolitan city. There is now another movement to save the wetlands of south-west Calcutta. Various public interest, environmental and people's science groups as well as the Left Front state government has taken note of the plans of the Calcutta Port Trust (CPT) to destroy the wetlands of south-west Calcutta. The CPT has been working on its plan to seize the wetlands and convert them into real estate. It has initiated action against hundreds of fishermen and their representative body, the 'Mudially Fishermen's Cooperative Society Limited' (MFCS). The south-west wetland movement is not only an attempt to save the detoxifier of the city's biosphere, but is also a struggle for survival and for the rights of the dalits and their cooperative.

The wetlands of Calcutta are one of the largest of their kind in the world, spread over thousands of hectares. They have a tradition, more than a hundred years old, of disposal and utilization of urban waste for agriculture and fisheries. They largely involve a comprehensive set of folk technologies, which demonstrate alternative and sustainable ways of recycling waste, without calling for multinational bank assistance programme or the use of expensive treatment technology.

The city of Calcutta slopes east and south, away from the river Hooghly, towards the sea. The wetlands to the east are a dumping ground for the city's waste. Also, 800 million litres of the city's sewage

Published in *Frontline*, 20 November 1992 and *Economic and Political Weekly*, 4 December 1993.

is canalled across it to the river Kulti en route to the sea. This wetland covered about 77 square kilometres in the 1950s. About 30,000 people worked in recycling Calcutta's waste, converted into manure for farm fish, vegetables and paddy. In this process 30 to 40 per cent of the sewage also got purified. Though the area is much depleted now, more than 20,000 work here to recycle the sewage to produce about 20 per cent of the city's requirement of fish and vegetables.

The area where waste-use farming was predominant was named the Waste Recycling Zone. On these waste and garbage farms, uniquely planned with alternative bands of garbage-filled lands and channel ponds, 150 tonnes of vegetables were grown daily. The ponds were filled with sewage twice a year, and after a sufficient detention time to allow for natural treatment of the sewage, it was used for irrigation of the vegetable crop. The same ponds were also used for commercial fish farming. This system, which probably started around the turn of the century, was amongst the first sewage-fed fisheries in the world. This part of the zone was called the Dhapa Square Mile. The second part was the fish pond region, locally called *bhery*, consisting of a grid-like network of channels to supply untreated sewage at one end and to drain out the used and excess water at the other end. Thousands of fisherfolk work in these ponds, supplying about 20 tonnes of fish daily to the city. The third part of the zone consists of thousands of acres of fields which use the effluent from these fish ponds to grow paddy twice a year. The wetlands thus perform for the city the functions of kidney and liver, and also those of the lungs, helping recharge the oxygen, clean the air and maintain pleasant weather.

A report prepared by the World Wide Fund for Nature-India on the wetlands of Calcutta is highly appreciative of the wisdom of this traditional practice. For example, solar radiation in India, of the order of 250–600 langleys a day, is always adequate for photosynthesis. The sewage ponds act as solar reactors. Solar energy is tapped by a dense population of plankton which also plays a highly significant role in degrading the organic matter. The excess plankton is consumed by the fish. The fish thus play a twin ecological role—they maintain a proper balance of the plankton population in the pond and also convert the available nutrients in the waste water into a readily consumable form for human beings. The report noted: 'What however is more striking is how easily these complex ecological processes have been adopted by the poor farmers of the wetlands of Calcutta. These 'natural ecologists' have developed such a mastery of these resource recovery activities that they are easily growing fish at a yield rate and production cost which is unmatched in any other freshwater fish ponds of the country.'

The wetland area, known as Mudially, was previously a part of the Garden Reach municipality. In 1983, it was merged with the Calcutta Municipal Corporation. The wetlands and the adjoining areas belong to the Calcutta Port Trust (CPT). According to Dhrubajyoti Ghosh of the Calcutta Metropolitan Water and Sanitation Authority, the wetland area was used as a site for solid waste disposal by the CPT and the CMC as recently as 1989. In the process, a part of the area was filled up. Apart from dumping solid waste into the wetlands, the CPT did not put them to any productive use. The local inhabitants used to catch fish from these wetlands and some also practised fish culture on a very small scale. During the mid-1950s, the CPT granted fishing rights on about 54 hectares of these wetlands. About fifteen fishermen from the western districts of Howrah and from Hooghly were employed for cleaning the water bodies and for culturing fish. They were engaged in various subsistence activities such as market gardening, horticulture and pisciculture. Some of them were on contract as fisheries labourers in the Department of Fisheries, Government of West Bengal. During 1957–8, some of them formed a group and obtained fishing rights to some small ponds in the area and began fish culture. In 1961, with the help and guidance of the Department of Fisheries, they formed the MFCS, a unique experiment in the use and development of wetlands.

Led by a group of seven fishermen, the MFCS started with a deposit of 25 paise a day per member in the 1950s. The cooperative society was registered in November 1961. From there it grew tremendously in terms of membership, activities, production, and profit. However in the past few years, the members were in constant fear of losing their means of livelihood. First, the CPT started curtailing the areas of wetlands and distributing portions of them to the dock authorities. Besides, industrial waste and silt from the river Bhagirathi made their work difficult.

Meanwhile, the society had developed a completely indigenous bio-engineering system to perform three important functions. The waste water was used as an input to grow fish. Its quality was improved before releasing it into the Ganga. An ecologically balanced system was developed to accommodate a number of animal and plant species.

In 1985, a young, committed and imaginative official, Mukut Roy

Chowdhry of the West Bengal Fisheries Department, joined the society and helped to evolve a scientific system to use the tanks and to develop intensive fish culture. The turnover of the society jumped from Rs 18.62 lakh in 1985–6 to Rs 37.38 lakh in 1986–7.

The technical system developed by MFCS for the treatment of waste water is unique. The average daily loading of sewage water is approximately 23 million litres, of which about 70 per cent is from industries and the rest from domestic sewage. This sewage initially passes through the first of six ponds. In this tank, called the anaerobic tank, the water is treated manually, using either lime or a biochemical. Water hyacinth is usually kept near to facilitate the absorption of the oil and grease in the effluent. Often the tank is dug in order to reduce sludge deposition.

The second tank, connected to the first through a narrow passage, is the breeding ground for hardy 'exotic' fish. The water is then allowed to flow into the third pond. The quality of water improves in each succeeding tank, which permits better utilization of the recycled nutrients and mineral contents of the sewage in fish culture. Finally, the water is let into the Manikhal canal system which eventually joins the Hooghly.

The society has developed a laboratory at its Nature Park. Here, all MCFS members are trained to test the quality of the water. When the effluents released are found to be deadly, there is a provision to take the water through the entire cycle again to improve its quality. A green alga called uglina performs photosynthesis and generates nutrients. As a result MFCS does not need to apply supplementary food except in the rainy season when photosynthesis becomes uncertain.

Fourteen dominant types of fish fauna have been identified in these ponds: twenty weid type fish fauna, three prawn and crab fauna, three small fauna, one spounge fauna and two snake fauna. MFCS is also successfully growing prawns. In early 1991, three sensitive fresh water varieties of fish (Chanda, Mourala and Punti) were found in some of the ponds, indicating an improvement of the quality of water.

The society produces about one tonne of fish a day. Its gross profit has increased from 20.52 per cent in 1985–6 to 50.90 per cent in 1990–1. Following a consciously adopted policy of creating maximum employment for casual labourers, in 1990–1 the society created 95,000 persondays of casual employment.

The society has also undertaken a massive afforestation programme, planting nearly 100,000 saplings of 78 species, out of which

60 per cent survived. Thirty per cent of the plants were from the leguminous group (like Subabulu), 30 per cent dust-and-chemical-absorbing plants like Akand and Neem, 30 per cent plants for attracting birds and 10 per cent horticulture plants. Leguminous plants were planted close to the water because the leaves falling in water become supplementary nutrients for fish culture. Bird-attracting plants like Trifala were planted far away from the water bodies. A study done by the Zoological Survey of India in 1991 disclosed that as many as 120 varieties of birds were observed in the Nature Park, 27 of them migratory. The society successfully also promoted spotted deer, duck and tortoises in the nature park. (The number of deer increased from 12 to 24 in 1990.) Thus a dumping ground of urban refuge has now become a sight for sore eyes, attracting many visitors everyday. Its contributions have also been recognized in terms of the awards it has received, including the National Productivity Council award and the Agri–Horticultural Society of India award.

The MFCS is also engaged in the sale of fishery inputs and consumer goods. Currently, it is building up durable consumer retailing through 80 enlisted retailers who sell in 37 markets of the city and buy fish from the MFCS at the usual 10 per cent discount. The retailers are nominal members of the society. They get incentives for higher transactions. The society is already selling processed fish in polythene packets to selected retail stalls. It also has plans to utilize the household labour of the womenfolk in the households of its members on a much larger scale.

Thanks to the enriched micronutrients available from treated water and the other steps, the cost of production of fishes by the MFCS has steadily declined over time. Among the society's welfare activities for its members are full reimbursement of educational expenses up to the primary level for children, a grant of Rs 2000 for the marriage of a daughter, and a sum of Rs 1000 as funeral expenses when a member dies.

The society has a three-tier membership structure, with absolute equality of members within each tier with respect to workload, pay structure, and other facilities. From the chairman to a new member, everybody gets a monthly salary of Rs 1500. To uphold the principle of equality in terms of workload and to avoid the problems associated with specialization and hierarchy, every person is required to do work in rotation. This discourages anyone from thinking that he is indispensable. A disciplined approach is adopted to enable a division of the total workload into various departments. The society is rigorously strict with respect to the fulfilment of duties by members while at work.

In 1985–6, the society received a loan of Rs 12 crore with a 50 per cent subsidy from the state government's FFDA scheme to enable the reclamation of 65 acres of wetlands under its jurisdiction. However, the members soon felt that the work done under the supervision of state government officials was not consistent with the society's design and requirements. As one recent study on the MFCS by Samar Dutta and Sanjeev Kapoor of the Indian Institute of Management, Ahmedabad puts it: 'While the whole world has been after IMF World Bank loans for restructuring and development, this small society without any financial aid from any quarter has thrived on poisons ... WCS has evolved a resource-recovery practice which provides a tutorial, ecosystem for the rest of the world, especially to the low-lying cities in the tropical climate regions like Dhaka, Bombay, Jakarta and Bangkok.'

These wetlands are situated in the industrial south-western part of the CMC area and to the south of the river Hooghly. Here a large area of land has been leased out by the CPT to different industries depending up on import/export through the port. Thus areas adjacent to the Kidderpore and Netaji Subhas docks have a number of chemical industries, repair shops, petroleum storage tanks, godowns, engineering industries and so on, forming one of the largest industrial estates of Calcutta.

Now the CPT has categorically spelt out its intention of filling up all the remaining 200 acres of the wetlands in the Taratola–Hyde Road in the industrial area of Calcutta. All along, the port authority has been filling up these wetlands in phases. In 1991, it tried to increase the lease amount enormously. It offered a licence to the MFCS for three years, one year at a time, on an annual fee of Rs 87,891 for the first year, with a 25 per cent increase every year. The society refused to pay the increased fee and asked for relief. The CPT stated that it needed to fill up the wetlands for the extension of the dock, container park, container repair yard, truck terminals, warehouses, water basin facilities, and so on, not to mention the construction of a 100 foot road and housing for security personnel and others. Using these pretexts, the CPT asked the MFCS to vacate the land and water body through a notice dated 15 July 1992. On 23 July 1992 it also decided to take possession of the land and water body.

The state government intervened in favour of the MFCS. It promulgated Section 144 in the area and stopped the CPT from taking action. The Fisheries Department even proposed to take hold of the wetland from the CPT, but the CPT refused the request for transfer of the water body and the land. In an effort to dislodge the cooperative society, the CPT released the findings of a self-commissioned NEERI report. The CPT stated that according to the NEERI study, the fish produced by the MFCS is unfit for human consumption. Nevertheless, the full text of the NEERI study revealed that although toxic elements were present in the initial ponds, the toxicity was within limits after the unique treatment method. The study further noted how it could be concluded from the waste water analytical data that the characteristics were indicative of the fact that the wastes originated from industries/institutions and were not purely domestic in nature. Organic and nutrient concentrations were therefore limited. On the other hand there were toxic elements, though within limits. The use of *jheel* water for bathing, swimming, and recreation necessitating physical contact was not recommended. Using treated water as water source would improve the water quality considerably. It was, however, recommended that the jheel water might be restricted to the use of wildlife propagation/irrigation and as a source of water for industrial cooling, fire fighting, and aesthetic development.

Dr A.C. Roy, Chairman of the CPT commented that it was not desirable that pisciculture continued in these jheels. In two areas of the CPT a licence had been given to the MFCS for fishing. One was the water area in Dhobitalao which is across the Circular Garden Reach Road in continuation of the land in Netaji Subhas Dock. The other area was on Taratala Road. A part of the Dhobitalao area was now in use as container parking area and container repair yards for the port. This was in intensive use and the CPT was under pressure to extend the area for the container park and also for repairs of containers, he said. Roy alleged that there were a large number of encroachments along both the jheels. To prevent encroachments, it was absolutely necessary to have CPT security guards in different areas of the CPT land. The MFCS was forcibly and illegally preventing the security organization of the CPT from carrying out these tasks.

Sujit Banerji, Secretary, Department of Fisheries, Government of West Bengal, disagrees. According to him, the society has done good work not only in developing pisciculture through waste recycling but also setting up a nature park. All these activities have substantially contributed to improving the quality of the environment in the area. The CPT itself appreciated the work and it was at its instance that the present executive officer of the cooperative society was retained by the Department of Fisheries to continue the work done. Banerji strongly recommended that in the present context of indiscriminate filling up of water areas in the Calcutta metropolis, the existing water areas engaged in pisciculture should be preserved by all means not only for fish production but also for improving the quality of the environment. People's science groups, environmental groups, labour and health movement groups like the Nagrik Manch, the Scientific Workers' Forum, the Vigyan Vikas, the Institute of Engineers, the People's Science Coordination Centre, *inter alia* have jointly taken up the challenge to confront the CPT. The state government is on their side.

POSTSCRIPT

The movement to save the wetlands of Calcutta achieved significant success in 1993. The West Bengal Fisheries (Amendment) Bill 1993; the historic decision of the Calcutta High Court; the People's Convention organized by the coordinating body of thirty-six organizations regarding the MFCS; the Centre's clearance of the Rs 132 crore action plan submitted by the West Bengal government to conserve the East Calcutta wetlands, and many other developments, signify the singular strength of people's movement, which has shaken various organs of the state on this particular issue.

The West Bengal Fisheries (Amendment) Bill 1993 adopted by the West Bengal Assembly legislation strictly prevents the filling up of wetlands that exceed 0.035 hectare in the state's notified urban areas. The State Fisheries Minister Kironmoy Nanda stated that the legislation aimed to help fisherfolk to set up fishing cooperatives in the wetlands. Those wetlands which had been illegally filled up in the recent years by real estate promoters would be confiscated, restored to their original state, and handed over to the fishermen's cooperatives.

This legislative initiative was inspired by the judicial intervention by the Calcutta High Court. Justice Umesh Chandra Banerjee in the case of MFCS, Nagrik Manch and others vs State of West Bengal, The Calcutta Port Trust and others, stated, 'Wetlands should be preserved and no interference or reclamation should be permitted ... . There shall be an order of injunction restraining the State Respondents from reclaiming any further wetland. There shall also be an order of injunction prohibiting the respondents from granting any permission to any person whatsoever for the purpose of changing the use of the land from agricultural to residential or commercial in the area as indicated in the map. The State-Respondents are further directed to maintain the nature and character of the wetlands in their present form and to stop all encroachment of the wetland area. The State-Respondents are further directed to take steps so as to stop private alienation and, if required, by extending the statutory provisions in regard thereto.' Regarding the Calcutta wetlands, the judgement said, 'There are 40 species of algae and two species of fern, seven species of monoceds and 21 species of diceds. Latest data suggests presence of about 155 species of summer birds of which 64 species are resident birds and 91 are migratory. These migratory birds are mainly from Siberia and East Europe and they arrive at the city through Trans-Asia Migration Route. Admittedly, Calcutta has had around 20,000 acres approximately of wetland area, of which 10,000 acres have already been reclaimed.'

It further said, 'There cannot be any manner of doubt that the Calcutta wetlands present a unique eco-system apart from the materialistic benefit to the society at large. Within the Calcutta metropolitan area, the Calcutta wetlands can be easily identified as the most outstanding wetland cluster ... . These wetlands bear the oldest tradition in the world of resource recovery from city's waste besides being one of the largest of such systems in the world.'

The judgement also added, 'Satellite township is a modern phenomenon in this country. Delhi has experienced few years back the same and so has Bombay and Madras. It is nothing new that this mad rush continues throughout the country for urban area. But does that mean and imply that the gift of nature to humanity shall have to be destroyed? Does that mean and imply that this unseen storehouse of nature's bounty would have to be exploited to its optimum level? In my view, the answer cannot but be in the negative. Wetland acts as a benefactor to the society and there cannot be any manner of doubt in regard thereto and as such encroachment thereof would be detrimental to the society which the law courts cannot permit. This benefit to the society cannot be weighed on mathematical nicety so as to take note of the requirement of the society—what is required today may not be a relevant consideration in the immediate future.'

Against this background, the Centre has also cleared the Rs 132 crore action plan of the West Bengal government to conserve the East Calcutta wetlands. According to an official spokesman, the Calcutta Metropolitan Development Authority had prepared a detailed management action plan, termed the 'Waste as Resource' project, which was submitted by Chief Minister Jyoti Basu to the Centre a long time ago. The project being undertaken jointly by the Centre, the state government and the conservation authorities provides for a self-reliant waste water treatment facility for Calcutta, increased productivity and ecological stability of resource recovery practices, augmentation of work opportunities for the rural and semi-urban people, and enrichment of bio-diversity. The project will regulate flow of sewage from the city into the sewage-fed 'bheris'. The ageing canal will be restored and upgraded. A subsidiary drainage system will be taken up and an open-air zoo, water sports facility and a bio-diversity park will be set up in various phases.

In spite of these positive developments, the people's movement to save the wetlands, under the coordinating body of thirty-six organizations, does not want to rest. A people's convention was organized in which more than five hundred representatives of various organizations participated. The convention noted that there was a deliberate attempt on the part of the CPT to unsettle the life of poor families and also to not accept the legitimate demands of the fish farmers for a long-term and reasonable rent for the land. The CPT also declared openly that the embankment and water body within the Nature Park, measuring about 80 hectares in two parts, would be used for the expansion and development of Calcutta Port.

Forests under Siege

People versus forests

Conservation under Tinsimani Sarkar

Asking for Wood makes You a Naxal

The flags, hoisted on poles, can be seen from a distance. They signify victory for the inhabitants of Palia Pipariya village, in Vankheri tehsil at the eastern end of Hoshangabad district, Madhya Pradesh against the high-handedness of the Forest Department. The story of the poles and flags began with Kubbu, a landless tribal labourer. He had bought eleven poles to repair his roof. On 20 September 1989, Forest Department guards came, confiscated the poles and asked him to pay a fine of Rs 1000 if he wanted them. Since he had no money, they took away the poles. A rich landlord of the village paid Rs 300 to the guards, and they handed over all the poles to him.

For the villagers, the issue of Kubbu's poles and his harassment became a question of the dignity and rights of all of them. They assembled that night and vowed not to allow anyone to remove the poles from the village. When the police, the Forest Department guards and the landlord's men came to take away the wood on a lorry, the villagers surrounded the wood, and the men had to return empty-handed. The villagers shifted the poles to the nine different tolas of the village, and hoisted on them flags of Kisan Mazdoor Sangathan, a mass organization of tribals and labourers in the area.

Kubbu said 'My respect has been restored. My harassment has been taken care of. This is more than sufficient for me.' His house was later repaired with the wood borrowed from a neighbour's house. On the night of the incident the villagers wrote to the Deputy Ranger, 'Dear Sir, today, 20 September 1989, you entered the house of an adivasi, Kubbu and confiscated his wood, and threatened him at gun point to pay a fine. This action of yours shows a complete disregard for our way of life and dignity on your part. Some wood is kept in every household for domestic use. If your laws say that it must be

This is a substantially revised version of the collection of articles originally published in *Navbharat Times*, 23–5 February 1990.

confiscated, then we all are culprits. So, would you accuse all of us?'
After this, the district administration sent notices to eight villagers,
but they did not accept them. The flags still fly aloft, after two and a
half years.

Palia Pipariya has been a centre of activity since 1980–1, after a
voluntary organization, Kishore Bharati, came to the area and gave
some initial impetus to the villagers to unite. In 1981, they, along with
Kishore Bharati, took the initiative to stop the mismanagement in the
distribution of sugar by the ration shops. This was how the Sangathan
came into being. The landlords strongly opposed it. Their men even
physically attacked the active villagers, and many were injured. The
malpractices were later stopped to some extent, and the poor began
to get their quota of sugar.

Once an organization took shape, there were various activities in
many directions. More than a hundred acres of government land in
the village was under the control of landlords and Patels since long.
The Sangathan demanded the measurement of government land and
an end to its misuse. When nothing came of it, they decided to use the
remaining land, around fifty acres, for collective farming. Soon after
the government decided to set up a silk centre there and the process
of vacating the other occupied lands started. Many employment ave-
nues have been created by the silk centre, not only for Palia Pipariya,
but for other villages as well.

✛

Hoshangabad district, among the Satpura hills, is home to the Gond
and Koraku tribals. Geographically, the district is divided into the
eastern and western halves. The east comprises Sohagpur and Hoshangabad tehsils and the west of Siwani-Malwa and Harda tehsils. The
western half, lying between the Satpura hills and Narmada river
looks fertile and productive, whereas Sohagpur is dry and dusty. The
eastern half is also more poverty-stricken than its western counter-
part. Similarly, the plains area of the western half has greater access
to roads, schools, and irrigation facilities than the hilly areas of the
east. The valleys of Vindhya and Satpura are well known for their
forests, which abound in teak. Sal (*Shorea robusta*), sisam (*Dalbergia
sisu*), mahua (*Bassia latifolia*), jamun (*Eugenia jambolana*), black myro-
balan, bamboo, and tendu leaves are also found.

According to a survey done by Kishore Bharati in the tribal village
of Rorighat around 20 kilometres away from the Pachmarhi tourist

centre, all the thirty-two households own 0.5–1.5 acres of land. In village Kharshali, out of a total of 75 households, 15 were landless, 22 had less than 5 acres and 28 had 5–12 acres of agricultural land. Among them, 10 households owned the bulk of the agricultural land (45 per cent) in the village, with one household owning more than 80 acres.

A tribal of Rorighat village earns an average of Rs 526 for 100 days of work. In a year, he spends 33.6 per cent of his time in agricultural work, but gets only 17.8 per cent income out of it. Wage labour consumes 41.6 per cent of time and gives 48.8 per cent of income. The average daily wage is Rs 5.24. The occupations based on forest-based produce generate 33.4 per cent of income in 24.7 per cent of the time.

It is easy to see how depriving the tribal people of the use of the forest resources can affect their survival. They do not get sufficient wood for *nistar*, i.e. wood for house construction, fuel, and plough. *Ranawa*, i.e. licence and fees to graze animals in the forests has deprived them of an age-old right. Displacement induced by many governmental projects has made their life hell. The plantation being put up by the Forest Department in Sardeh village has deep pits with big enclosures within and outside the village covering the villagers' residential and agricultural land. Notice boards warn the villagers not to enter or let their animals enter these plantations. Wrongful entry will cost them a fine of Rs 500 and their cattle is impounded. In Nazarpur village, the Forest Department has taken over 35 acres of agricultural land of 15 families for plantation, and the villagers claimed not to have received any compensation. Grazing lands are also being converted into plantations. Under ranawa which has become in vogue in Madhya Pradesh since 1986, the rates of grazing are Rs 10 per year for a cow, goat, or sheep, and Rs 50 for a camel.

✢

In most villages of the plains and the hilly area of Hoshangabad district, the contest over wood has become bitter and violent. The villagers needed the wood from the forests regularly, and they collected it either on the sly or by giving some money to the forest employees. Whenever forest guards were absent from the fields, the villagers used to cut and collect some wood. The forest guards warned them at intervals. In September 1991, matters boiled over in Kursikhapa village, 25 kilometres from Piparia. Ten persons of the village were arrested on charges of theft and for their row with the forest employees. It was alleged that at the Matkuli Range Office, the villagers were hung upside down, given a good beating, and later sent to jail. A guard, on whose complaint the villagers were tortured, was caught after some days for possessing stolen wood and was suspended.

In Dakrikhera village, 10 kilometres from Kursikhapa, Rangalal's two sons were arrested on charges of stealing wood. After his sons were released on bail, all the male members were called to Matkuli office and detained there. The same night, the guard and the Forest Department employees entered their houses and beat up their wives and children.

On 13 December 1989, a flying squad of the Forest Department came to Junheta village on the outskirts of Vankheri Subdivision, accompanied by armed policemen. They entered the houses and pulled out several poles from the roofs, damaging many houses. The menfolk being out in the fields, the women and the children were beaten up. Later some men of the village and two activists of the Kisan Majdoor Sangathan were arrested, charged with rioting, threatening to kill policemen, and obstructing them from discharging their official duties. Hundreds of people from Junheta, Palia Piparia, Nindwara and other areas assembled at Vankheri that night to protest against this incident. They staged a dharna and demonstrated in front of the Ranger's office. A big demonstration and a public meeting were held a week later at Piparia. Realizing the intensity of the people's anger and agitation, the administration ordered the transfer of some officers involved and payment of compensation to the villagers whose houses were destroyed.

In 1987, Salai, a tribal village of thirty houses in Siwani tehsil underwent a severe thrashing over a small bit of wood, and many villagers faced serious criminal charges. One of the villagers, Channu, was in the forest taking some wood from a tree for his cot. A *nakedar* saw him, but did not do anything. After two days, two of them came to the village and asked money from Channu. When he refused, they started beating him and ordered him to accompany them to the forest office. Some villagers came and tried to pacify the *nakedars*, but when this did not work out, the villagers forcibly freed him. The next day forest guards came, beat up many people including Channu and his father, looted houses, misbehaved with the women, and arrested eight villagers.

Some walls of Salai village carried graffiti, uncommon for a remote tribal village: 'Bamboo poles, wood for domestic use and fodder have been our right since ages'; 'If the jungles are ours, there will be peace or else there will be violence'; 'The jungle belongs to all, it is not anyone's personal property'.

+

Impelled by the Kisan Adivasi Sangathan, the villages located in the forests of Kesla–Bhora region of Hoshangabad are campaigning to retrieve the money that the administration, police and employees of the Forest Department have taken as bribes from the villagers. The Sangathan asked the villagers not to give money to any government officer at any level. They were urged to report about anyone found accepting a bribe, and keep fear out of their minds. Many villagers were emboldened to act. The villagers of Banjaridhal submitted a petition saying that the *Gram Sevak* Malviya passed cases under IRDP for them, but extorted money: Rs 200 from Channu Lal, Rs 800 from Ramji Lal, Rs 200 from Vatan Singh, Rs 650 from Sukru Lal and Rs 360 from Ram Singh. Shyamu showed a letter of appeal: 'We are the inhabitants of Doodar forest village. Thanedar Pana Beejadehi accused us tribals of unearthing hidden wealth from a well, and after threatening us, took Rs 1000 from each of us.' Lakshman's letter from Goli village said that a Guard and Deputy Ranger of Medakhera took Rs 120 from him as pasture tax. According to Raj Narayan, from the Kesla region alone, the Forest Department, police, the Revenue Department and the bank, on an average, collect about Rs 500–1000 annually from each family as bribes and fines. Cash, mahua, foodgrains, milk and ghee—there are various ways of this extraction. In this way, annually, between Rs 50 lakhs and one crore is collected illegally from about 10,000 households of the Kesla sub-division.

When corruption continued unchecked and their complaints were not heeded, the villagers resorted to agitation. In November–December 1989, hundreds of villagers from Bhora village staged a dharna in front of the Forest Range Office. Many other campaign programmes followed this. After this, in the presence of the District Forest Officer, the deputy ranger returned Rs 1150, taken as bribe from five tribals. In similar victories, Kaliram received Rs 1500 from the sub-tehsildar of Chindapani village under Kesla. The Patwari of Marwapura village also returned Rs 1500 to six tribals. The Patwari and Nayab Tehsildar had taken away Rs 27,500 from four tribals of Chindapani who had received the money as compensation for their land from the public sector NTPC. The money was returned. After settling many such cases, the tribals went further. An indefinite dharna was staged in front of the Bhora Range Office, which lasted for three days. The administration capitulated on it. In a settlement that the villagers negotiated as equals with the district administration and the Forest Department, it was decided, among other things, to appoint a special officer to look into public complaints. This still worked, and seemed to be percolating to the Pipariya and Vankheri regions as well.

Meet the Forest Minister
of Tinsimani Sarkar

Meet Bhiku Urao, labourer and Forest Minister of Tinsimani Government in Mahuwa toli of Chotanagpur, Bihar. With six other ministers and a prime minister, he is the member of a cabinet of a government, formed in 1989 by the tribal villages of Vishunpur region, to look after their forest, water, and land. The government, which has its own 22-point programme, meets for two hours every week. '*Simani* means a place and *tin* means three. We had a tradition in this region of three villages joining hands and sitting together at one place to decide their future plans. Now the tradition has been extended to the entire region', explained Bhiku Urao. Though not a parallel government, unless the state government recognizes their existence and also does something to resolve their issues, they will continue to boycott the elections, political parties and the administration. The Tinsimani Sarkar has called for a total ban of wood and forest cutting here, and alleged that the state government is involved in illegal felling. The latter has denied it and held the tribals responsible for breaching the peace in the region and not allowing the administration to perform its essential works. Hundreds of tribals in dozens of villages are facing many court cases. The government wood depots are closed at several places.

Primarily inhabited by Urao, Munda, Asur, Ho, Gond, and Virajia tribes, the Vishnupur region, 123 kilometres north-west from Ranchi town, hosts many important agencies of the forest department. For the tribals, this has meant increased exploitation. The villagers complain that a few years ago, the Forest Department used to give them rights in the forests near the village so that their requirements of woods were met. Now they either got small concessions far away from the village or none at all. They were stopped from taking firewood from the forests, were also regularly arrested, jailed, and implicated in various cases.

The Forest Department auctioned the trees to contractors for felling. The indignant villagers could do nothing about it, but now they can at least go to their sarkar, which listens to them. The sarkar held meetings in village after village and formed 'save the forest committees'. A meeting organized by the sarkar in Heshrag village in 1990 took a blanket decision not to allow any cutting and selling of the forests in this region. A traditional song of Uraos became the slogan of the government: *Mati katu jungle, Mati katu pahar, Jungle mein sona bhair gel la, Adivasis ho jungle bachau, Hamar raij ke hariyer karu.* (Do not cut forests. Do not cut mountains. Forests are full of gold. Oh, tribal brothers, Save the forests, Keep our kingdom green).

The villagers put locks on the wood depots of the State Trading Corporation of the Forest Department, told the forest employees to leave the depots, and persuaded the labourers from outside not to come for cutting. They jammed the roads, laying siege of trucks loaded with wood and gheraoed the forest officials.

The resistance began at Heshrag village. The village men and women gathered in large numbers one morning at a point of the forest where the contractor was supposed to start cutting trees. They seized the wood-cutters' tools and told them to leave. The wood-cutters left, but returned after a few hours with the contractor and some forest guards. After some altercation between the labourers and the villagers both party left. The next day, police came and arrested ten villagers, threatening to arrest the entire village.

The most difficult thing for the resisters was to persuade the outsiders and even some villagers not to go as wage labour for forest cutting. The Forest Department was offering an attractive wage of Rs 17.50. In some villages, the officials offered a bribe of Rs 1000 to start cutting the trees, which the villagers declined.

After the success of this resistance movement the Tinsimani Sarkar decided, in 1992, to go and capture the waste and governmental lands in the villages, and grow plantations on them. Plants of mango, jack-fruit, pronun, tamarind and mahua could be seen in these captured plots. Mulberry dominated, because the villagers wanted to develop their silk industry.

While the rules formulated by the sarkar and the village committees for the conservation of forests differed from village to village depending upon the local situation, some of the broad principles followed by all were: the trees would be cut only for house making and fuel needs; small, raw or sal trees would never be cut; bamboo seeds would not be used for pickles; forest fires would be put out immediately and nobody would light a fire within the forests, etc. Violators were fined seven rupees, of which two went to the informer and the rest to the forest committee.

Trees that Know, Bleed and Talk

KHAKSITOLI: HE SAYS HE HAS THE SUN

A traditional king who lives in a tola and rules a tiny tribal community/has a small territory. Under his command, a government forest on 37 acres of land has been saved for the last thirty-five years. He says he has the sun. The Khaksitoli of the sixty year old *Parha Raja* Simon Oraon is some 45 kilometres from Ranchi town on the Ranchi–Gumla–Mumbai highway, and then five–six kilometres walk from the highway. The raja may be found somewhere in the remote forests, or labouring on his three acres of land. He has the last word on issues concerning land, forest, water, property, marriage, and festival in *barahpara*, a tribal panchayat of twelve tolas. His toli still obeys him, and the dense forest in Khaksitoli, comprising sal, mango, blackberries, and kathal, reflects this fact.

The forty-six Oraon families of Khaksitoli believe that the trees in their environs belong to them as a part of their family and village. From the trees comes wisdom, knowledge, and life. Following the nationalization of forests, when leases were given to contractors and the felling of trees was stepped up, the villagers themselves felled a lot of trees on community and private lands. As Simon Oraon put it, ' The villagers suspected that forests taken over by the government would bring them harm. The hard fact of the loss of ownership over forests has as not yet been accepted here, and probably will never be.'

The villagers soon began to miss the trees, which had always been there as members of their family. The Parha Raja took the initiative and summoned the panchayat of his ten tolas. It was decided that every village would reclaim its ownership rights over the surrounding forests from the government, and the contractors and each village

The collection of papers was originally published in *Navbharat Times*, 3–5 May 1990. This is a substantially revised version incorporating more recent developments.

panchayat would take up the responsibility of distributing forest produce among village families, after assessing their needs.

The implementation of the panchayat decision started in 1954–5 in all the 10 tolas. In every tola some persons were to guard the forests. Each family was to contribute grain to the panchayat, to pay these guards. But after three or four years, this arrangement broke down. Some stopped contributing the grain, the Forest Department harassed the locals by implicating them in cases, and sometimes the villagers fought over the distribution of forest produce.

Khaksitoli, however, adhered to the panchayat decision. In this base of Parha Raja and his panchayat, respect for traditional authority, gender, age and, to a lesser extent, riches are the main parameters. The Parha Raja has the authority to resolve every dispute except murder—the first day is reporting, the second day debating, and the third day his final judgement. Parha Raja could well claim that he had the sun, the source of all power.

In Khaksitoli, four persons were appointed guards of the forests. If anybody intruded into the forest to cut trees, they would beat a drum. The alerted villagers would rush in and order the intruder to leave quietly. Soon word spread that it was not easy to cut trees in Khaksitoli and even from the government forest.

In 1978, when the Forest Department leased out the Khaksitoli forest to a contractor, the people decided to resist. An appeal was made to other tolas of the Parha panchayat not to offer any help to the contractor. The villagers started guarding the forests with their bows and arrows. An arrest warrant was issued against Simon Oraon, but the contractor, his men, and even the policemen were reluctant to enter the village. After several months, the contractor was given some other forest area and Khaksitoli forest was categorized as 'not for sale'.

The next contractor obtained a lease from the Mining Department to quarry stones from the forest area and also obtained a no-objection certificate from the Forest Department. He also set up a stone crusher near Khaksitoli. Soon it transpired that the contractor's men were digging up stones all over the place and in the process uprooting trees and carrying them away. Having understood the contractor's real intention, the villagers, one day chased away his workers and tied the *munshi* to a tree. They seized the contractor's implements and fined him Rs 500. The police, on being informed, requested Simon to release the clerk while retaining the implements as guarantee for the payment of the fine. The villagers released the clerk but made it clear that the contractor could not enter Khaksitoli and its forest area. The contractor did not even return for his tools.

The traditional panchayat of Khaksitoli and its forest committee collected 20 kg of rice per year from every house to pay the guards. A person in need of wood for house construction or for agricultural implements had to present his case before the panchayat. The panchayat decided the actual requirement of wood, marked the particular tree or branch to be cut down. Without the permission of the panchayat, nobody could cut a tree or a branch even on his private land. Violators were fined severely. Anyone could pick fruits which fell on the ground but no one could pluck from the trees. On a particular day, fruits were plucked and distributed equally among the households. For the production of lac and tasar, ber, jujube tree, safflower (*Carthamus tinctorius*), dhak tree (*Butea frondosa*), and sal (*Shorea robusta*) were distributed among the villagers in turns.

The Parha Raja exhorted the villagers on the need to work hard in order to extract a living from the forest and the land. A depleting environment simply meant working harder and longer. One member of each household got together for some days in a month and worked on private or government wasteland. They did voluntary labour to construct small dams, nalas, and wells for storing rain water.

NAWADIH: MEN FOR MAHUA, WOMEN FOR ARJUN

The tribal women of Nawadih village in Chotanagpur region were adamant that they would not allow the planting of mahua trees. The officials and village men on the other hand were in favour of mahua trees. The women said that already there was large-scale distillation of liquor from mahua; the growth of more trees would increase liquor sale and consumption. The issue came before the panchayat, where after a long debate, the women's view prevailed. A list of tree species to be planted in 10 acres of community land was drawn up. It was also decided that each family would plant at least two fruit trees in its *baari*.

Next year, when plantation was to be taken again on nearly 15 acres of community land, the men and the women together worked out many new plans. This time, while clearing the bushes, several elderly villagers and other knowledgeable people were called to ensure that the medicinal plants were recognized and protected. The villagers decided to plant, apart from indigenous trees yielding fuel, fruit, oil, and medicines, some trees which could yield some cash income relatively early. They also decided, as an experiment, to raise a crop on one hectare of this land.

On two hectares, the silkworm-rearing Arjun trees were planted. One hectare of this was intercropped with *kurthi* (a pulse) and *sarguja* (oilseed), which enabled the villagers, for the first time to reap two quintals of kurthi and half a quintal of sarguja. The villagers decided to increase the crop from year to year. It was the same villagers who had initially rejected a proposal of the government officials to plant trees and protect them. When the officials showed willingness to revise their priorities keeping in view the needs and perceptions of the villagers, their participation in this task surpassed the officials' expectations.

In this hilly village, located about 60 kilometres from Ranchi, nearly 80 per cent of the families are Munda tribals. In 1982–3, some officials of the Forest Department and some other people wanted to start a project of tree planting on the community land, with the entire expenditure of the operation being borne by the government or other organizations. The villagers rejected the proposal, after discussing it for hours. During the summer, they said, there was a serious shortage of drinking water in the village. The Block Development Office was still digging a well after three years on the job. If the government or other outsiders wanted to help them, they said, they must first help to build a *pucca* well for drinking water. They were fully conscious of the need to protect trees, they said, and were giving full protection to the forests on nearly 20 acres of government land. They did not even allow the guards to enter this forest. They even unfurled a flag on the tallest tree to symbolize their rights over the forest.

It was therefore decided that before the tree planting work, a well would be dug. The villagers would contribute the labour and the Forest Department would pay for the construction materials and labour. In just about three weeks, in February 1983, the villagers had their well.

At this stage, the officials again broached the subject of planting trees on the community land. The villagers were willing, but what if the Forest Department took over the forest planted by them? The village rights over this forest had to continue, they said. They were willing to donate a part of their labour as free labour. It was finally decided that the village rights over this forest would be maintained. Every household would donate labour one day in a week, and on the remaining days, they would accept a lower rate of payment, i.e. Rs 7 for a day, instead of the government wage rate of Rs 10. To oversee the work, a joint committee of village men and women was set up.

A problem arose later in distributing the crop raised on the community land. Some of the better-off families had hardly contributed little to the labour involved in raising this crop, but as it had been raised on the community land, they wanted an equal share in the crop. It was decided in a meeting that all such work would be done by voluntary free labour, and the crop would be distributed equally among only those families who contributed labour.

Fruit trees—guava, jackfruit, roseapple, mango, jujube, etc.—overhanging with lush fruit welcome visitors to the village. The ripe fruit look ready to fall, but there is hardly any conflict over their ownership. The villagers are protecting the trees raised on the community land. Two households have been given this responsibility, and compensated for their contribution with grain, fruits and some cash by the villagers.

CHACHKOPI: 'PLENTY OF FURNITURE'

Chachkopi village strove to create a lust forest during the 1980s. When its efforts had borne fruit, it decided in the 1990s, quite democratically in a panchayat meeting, to destroy the forest. Today, in an area of around 40 acres, only remnants of cut trees remain.

Buddhu Urao in the village explained, 'It was sometime in February 1990. The village panchayat sat through to discuss the problem of continuing forest cutting, sometimes by the neighbouring villagers and sometimes by our own people. It came to the conclusion that if others were taking the benefits by cutting our trees, why should we not cut them ourselves, and reap the benefits. Early one morning, we all went to the forest to cut down the trees, and distributed the wood among ourselves. It gave us plenty of cots and furniture.'

Chachkopi, a village of Urao tribals, is in the Baro block of Chotanagpur region. It comprises 547 households spread over four tolas. Five Sahu families own 80 acres of land here, whereas Uraos have one to five acres of agricultural land. Two Urao households have no land. The village depends solely on agriculture for livelihood.

In the early 1980s, the villagers were finding it hard to meet their daily requirement of firewood, and wood for agricultural implements and housing. They had to either buy wood or steal it from the reserved forests.

A voluntary organization came to the village in 1978 and organized a village development camp. It conducted a survey, listing out the villagers' problems and solutions, according to their priority. During the course of this interaction, lasting more than three weeks, the village panchayat decided to create a forest within the village. In a widely attended meeting of the panchayat, it was decided to plant trees on panchayat lands, government lands, and private wastelands. The villagers volunteered their labour for levelling the land, digging the pits, watering the plants and guarding them. One guard was chosen among them, for a remuneration of 30 kilogram of rice per year. The voluntary organization and the Forest Department gave some areas for the plants, and fencing. In several phases, more than 40 acres of land was planted in a span of two years. The survival rate of the saplings was more than 75 per cent. When the villagers planted saplings on a plot of government wasteland that one of them had captured since long, he uprooted all the plants. The panchayat imposed a penalty of Rs 50 on him, instructed him to replant the whole plot, and even went ahead with the punishment. About five years later, however, the enthusiasm slackened.

The villagers virtually stopped giving money and grain for the protection of the forests, and the patrolling became indifferent. The voluntary organization and the forest department left the scene. The villagers cared more for their personal interests in distributing the forest produce, which bred ill-will. The neighbouring villages like Tuko, Bantoli, Pandepara, and Khukhura, which had no forests of their own, took advantage of this situation, and their people cut trees in the darkness. The panchayat assigned the night patrolling to the youth of the village, but once they were attacked, beaten and threatened by the poachers. Frightened, they stopped their patrolling. The panchayat lost its legitimacy in terms of forest management when it developed cold feet in taking action against some wood-cutters of the village who were supported by an influential Sahu family.

Once again the village has no forest and they have gone back to those poverty-stricken days when they ran around from one place to another for fuel, fodder, and wood.

Chronicles of Anger

MAHARASHTRA: THE BAMBOO BEAT

For many years now, Dandakaranya in the Gadchiroli district of Maharashtra, has been a troubled place.

The Thapar groups owns the Ballarpur Industries Limited at Ballarsha, Maharashtra, which produces one lakh tonnes of paper every year. The main supply of bamboo for this industry, plus many others, came from Dandarkaranya, in the Gadchiroli district of Maharashtra.

Casual labour during the bamboo felling season from October to June formed the main source of employment for the tribals of the area. Paid meagre wages—as low as 0.25 paisa for a bundle of *atvai* (felling) and *dolvai* (extraction)—they were trapped in a vicious circle of advance loans in exchange for a large amount of unpaid work. The rules required them to cut two metre-long bamboo pieces, but they were constantly forced to cut longer ones. If anyone by mistake cut a bamboo violating the silviculture method, s/he was beaten. Tribals past their prime were denied employment. Malaria, snake bite, sun stroke, and accidents at work were their common lot.

In 1983, the tribals organized themselves into the Dandakaranya Adivasi Kisan Mazdoor Sangathan (DAKMS), and agitated for higher wages. The management, having failed in its strike-breaking tactics, ultimately agreed to enhance the wages. As a result of the constant struggle of the DAKMS, the wages of the tribals increased to Rs 4.00 per bundle by 1992–3, together with a substantial improvement in their working conditions.

The lower employees of the paper mills, generally known as field staff, many of them tribals, also joined the bamboo labourers. They were divided into various categories like daily wagers, temporary, retentive, permanent, etc. Together they raised the demand in 1988 that retentives should be made permanent, which the management conceded after prolonged discussions, but did not implement the decision. After another encounter, the management made 41 tribal retentives permanent but left out 31. In 1989, a strike ensued, with the demand that all those who had put in ten years of service should be made permanent. Another strike followed in 1991, with the formation of a field staff union, and all those who had completed eight years of service were made retentives and those with five years service as retentives were made permanent.

The battle was waged also on other fronts. In 1989, the Forest Department started cutting bamboo in the Nistar Coops. The tribals successfully demanded better wages here as well. Thus, in the case of bundles of long bamboo pieces of 3.70 metres length, the payment per bundle was increased from Rs 70 to 90. Many problems still remain unresolved, and DAKMS is continuing its battle.

ORISSA: CONTROL OVER MINOR FOREST PRODUCE

Though minor forest produce (MFP) items provide a considerable revenue to the Orissa state government, the forest people, especially the women, who spend long hours in collecting it, were not allowed to stock, process or sell these items in the open market. The government leased out these rights to contractors. Even the Tribal Development Cooperative Cooperation (TDCC) usually sublet its lease to local contractors and businessmen, leaving the forest people vulnerable to exploitation.

The women foresters of Mandibisi, in the Rayagada district, decided to take a defiant step and formed the Mandibisi Mahila Mandal (MMM) in 1994 to wage a struggle for the right to collect and sell MFP. They started collecting hill brooms on their own. They had met the Chief Minister, Biju Patnaik, who had verbally agreed that they would be allowed to collect MFP. But the TDCC, though ostensibly working for the welfare of tribal foresters, was unwilling to relax its lease.

After various discussions and meetings which involved the panchayats, the government stated that the TDCC would give licences to registered women's groups as additional agents. They would collect and market the finished MFP items through the Orissa Rural Marketing Apex Society (ORMAS), which was ready to buy MMM's stock at Rs 7 per kg. The MMM agreed. In May 1995, however, the stock of brooms collected by the MMM was forcibly taken away by the Forest Department in the presence of officials of TDCC. Several petitions followed. In response to a public interest litigation, the High Court chose not to consider the rights of the forest women. Ultimately, the government stated that it was not possible to lease to MMM those MFP items which had been leased out to TDCC, without stating for how long the TDCC had the lease.

The explicit aim of the TDCC was stated to be ensuring fair price and facilitate marketing for forest products. In actual practice, however, the TDCC was achieving the opposite. The government constituted a price fixation committee at the district level every year to fix procurement prices of MFP items. The leaseholder agency had to pay these minimum prices to the collector, but surveys indicated that most MFP items were sold by the primary collectors at less than 50 per cent of the fixed procurement price. For hill brooms, the price received by the forest women collectors varied between one-third and one-half the procurement price. A series of other MFP products were contracted by private parties, which had a ten-year lease for twenty MFP items. Government policies were thus divesting the foresters from access to local resources.

The Divorce between Industry and Labour

Shramjivi Hospital, Howrah

Endless woes

Tomorrow's Theatre of Conflict
Pollution and Workers' Health

SLOW DEATH FROM STONE CRUSHING

'... A compensation of Rs one lakh each (which is still at the lower side) to the heirs of the deceased workmen would meet the ends of justice. We make it clear that the quantum of compensation being awarded in this case shall not be a precedent in other cases because the compensation in such cases should have been much more .... The first instalment of Rs two lakhs shall be deposited with the Labour Commissioner on January 1, 1997. The monthly instalments of Rs two lakhs shall be deposited thereafter by the 1st of every month till the total amount is paid. The Labour Commissioner shall proportionately disburse the instalment amount among the heirs of the deceased workmen.

We further direct the Labour Commissioner to have those workmen who are suffering from occupational diseases, medically examined within one month of the receipt of the order, so that the extent of disablement is determined ...'

— Kuldeep Singh and S.P. Kurdukar, Supreme Court of India,
Writ Petition (C) No. 3727 of 1985, 26 November 1996.

A lesser known judgement in the on-going case of *M.C. Mehta v Union of India*, popularly known as 'Ganga Matters', under article 32 of the Constitution of India, for the first time established a noble principle: a polluting industry damaged the environment outside the factory, simultaneously created havoc to the health of its workers inside the factory. Hence the industry had to pay the price of its pollution to the citizens in general, and to the sick and suffering workers in particular. Largely at the receiving end from the courts as

This is a substantially revised version of the paper originally published in *Economic and Political Weekly*, 30 August 1997.

well as the executive in several environmental litigations, workers at last were beginning to be recognized as an entity and taken seriously. The issue of workers' health and safety in a polluting industry which faced the threat of closure, assumed special significance. Workers were not only to be compensated for the loss or suspension of their employment. Their loss of health, their death or disability had also to be considered in the compensation package. This could also be with retrospective effect, especially in the case of polluting industries, whether closed or likely to be closed. The process by which the court judgement was achieved is as important as the judgement itself. This long process and its outcome gives a concrete example of the fact that environmental initiatives, workers' organizations and support groups, people's science activists and others can work together on the burning issue of industrial pollution, and can also promote their cause mutually. An activist in Calcutta commented that awareness regarding the interlinkages of worker's interests, occupational health and environmental protection was increasing, as reflected in some recent initiatives and legal interventions. This could be the main plank for joint actions in future.

The judgement pertains to the tribal village of Chinchurgeria, Jhargram in the Midnapore district of West Bengal. A stone crushing unit, Surendra Khanij Pvt. Ltd, started here in September 1987. The unit manufactured quartz powder from quartzite and supplied it to glass-manufacturing units. Quartzite was brought from the local forest areas.

Silicosis, a deadly occupational lung disease, is caused by the inhalation of free silica. The level of risk depends on three factors: concentration of dust in the atmosphere, percentage of free silica in dust and duration of exposure. Cough with sputum, decreasing body weight and general malaise are symptoms of silicosis. Chronic bronchitis is frequently associated with advanced silicosis which may progress to respiratory or cardiac failure.

In early 1993, the Quark Science Centre at Jhargram learnt through two foresters of the Forest Department of a series of workers' deaths and diseases in Chinchurgeria. Activists of the Science Centre went to the village with three doctors, and spent four days there. They tested more than a hundred villagers, several of whom were found to be suffering from silicosis. The villagers told them that they had gathered twice at the gate of the factory, demanding its closure, but the police came and threatened them, thus silencing them for good.

The Quark Science Centre now took the initiative to ensure the closure of the factory. First, it organized a massive signature campaign among the villagers and people of Jhargram demanding closure and compensation, and also many street corner meetings. When this failed, hundreds of villagers from Chinchurgeria and nearby villages assembled in Jhargram in April 1993 and sat on an indefinite dharna outside the SDO office. Thereafter, the SDO ordered the closure of the factory.

At this time, Nagrik Manch, a labour support group in Calcutta, along with six central trade unions—INTUC, AITUC, HMS, UTUC, UTUC (LS) and AICCTU—decided to take up this issue on a different plane. They forged a broader unity among different streams of trade unions to defend the case of the dead and suffering workers of a polluting industry, in consonance with the cause of environmental protection. Through the 'Ganga Matters', they all intervened in the Supreme Court and filed a public interest litigation, which read: 'The indifferent, irresponsible and self-centred attitude of the authority/owners of industrial units, which cause environmental pollution, is indeed responsible for the occupational diseases among workers of the respective units. That the workers are directly affected and suffer from such occupation specific diseases, would be, in part evident from the long list of unsettled claims submitted by the workers with Employees State Insurance Authority ... . Environmental pollution outside the unit and occupational diseases inside the unit are perhaps the two sides of the same coin. Whereas, the Supreme Court has already most justly and effectively passed a series of momentous verdicts regarding the "external" environmental pollution, we are keenly awaiting its verdicts and directives regarding the "internal" occupational hazards and diseases ...' They suggested to the Supreme Court that it should safeguard the right to livelihood of workers, besides its efforts at protecting the right to live. Besides, it needed to give directions to safeguard workers' health from occupational diseases, which was ultimately related to the factors creating environmental pollution outside the factory.

On 21 April 1995, the Supreme Court directed the West Bengal Pollution Control Board (WBPCB) to get in touch with the Nagrik Manch and other groups, and find out the details of workmen who were suffering from occupational diseases in the polluting industries. Accordingly, the WBPCB requested Nagrik Manch to submit a report on occupational diseases in West Bengal. The Manch, which is a non-funded, non-party, voluntary citizens' initiative on issues relating to labour, industry and environment, submitted a report. This report, the first of its kind in the state, identified some specific industries, including Surendra Khanij, where prevalence of occupational diseases had been inferred and reported. It was found that the workers of coal mines, stone crushing units, textile mills, jute mills, asbestos handling units, metro railway, municipal conservancy and sewerage work, and chemical industries in West Bengal, were suffering from several occupational diseases like silicosis, pneumoconiosis, byssinosis, asbestosis, hearing impairment, infections and other parasitic diseases. The WBPCB made further enquiries on the basis of this report, and found out that 20 workmen had died owing to occupational diseases in Surendra Khanij and 12 others were suffering from these diseases. It was further stated that the District Magistrate, Midnapore had also confirmed the death of a number of workers owing to environmental pollution from Surendra Khanij and some labourers suffering from occupational diseases.

Based upon the reports filed before it, the court gave two directions in February and March 1996. First, it directed the Secretary, Health, West Bengal and Director, Health Services, West Bengal, to render all possible help to the workmen suffering from occupational diseases who had worked at Surendra Khanij. Secondly, it directed the industry to reply to the court, giving its responses and comments to the facts brought out in the report of the WBPCB.

In the course of several hearings in the court, the management of Surendra Khanij and other polluting industries and even the West Bengal Government denied the existence of occupational diseases among their workers in unison. On 4 April 1996, V. Subramanian, Labour Secretary, Government of West Bengal, noted in his reply to the Supreme Court:

On inquiries being made by the Directorate of Factories, Government of West Bengal in the name of M/s Surendra Khanij Pvt. Ltd., it was learnt that the said unit was not registered with the Directorate of Factories. The said unit was visited by the Medical Inspector of Factories on 13.5.93, when the Unit was found to be closed. No information therefore could be gathered either from the Management or from the workers in respect of the matter. The Medical Inspector of Factories also visited the Jhargram Sadar Hospital for the purpose of further investigation. It is learnt that there was no report of admission of any such case in the said Hospital.

On 29 March 1996 the Supreme Court once again directed the Labour Commissioner to hold an inquiry to find out whether the workmen who had died or were suffering, had been working with Surendra Khanij. It also directed him to further indicate whether the workmen in fact suffered from occupational diseases as a result of the pollution generated by the stone crusher units.

Consequently, the Labour Commissioner appointed R.K. Saha, Joint Labour Commissioner, as an Inquiry Officer, who along with S. Ghosh, Assistant Environmental Engineer, WBPCB, made a detailed survey of the factory and the workers. Their report is revealing. The unit was not registered under the Factories Act, 1948, as required by law. It did not obtain any No Objection Certificate from the WBPCB. From the payment vouchers, it appeared that the daily wages at the beginning were Rs 12 for males and Rs 10 for females, which were subsequently raised to Rs 18 and Rs 16 respectively. Of the 20 deceased persons, 16 had worked in the unit. Since no records relating to their deaths were available, their actual cause of death could not be ascertained, but their relatives disclosed that all the deceased persons suffered from breathing troubles, weakness, etc. They were mostly engaged in jobs like sieving, filling bags with stone dust, loading and handling dust bags in a confined space, dust collecting from bag houses (which are not the conventional type of bag filters), etc. Most of them died even after treatment of anti-tuberculosis.

On 9 September 1996, the court asked the Labour Commissioner to assess the compensation in respect of both dead and suffering workers, after hearing both the parties. The Commissioner was also to determine the compensation to be paid to workers in other industries who had died of occupational diseases or who were suffering from these diseases.

It may be noted that Schedule III of the Workmen's Compensation Act specifies the diseases which are notifiable and for which compensation may be claimed. Schedule IV on the other hand provided for a table enumerating a list of factors, according to the age of the worker, which would be used to calculate the compensation. However, the rules and procedures of computation governing the payment of compensation to victims of occupational disease did not exist in the Act. Hence, scientific computation of compensation on the basis of Workmen's Compensation Act was not provided for in the instant case and so could not be completed under the Act.

At this juncture, the Nagrik Manch tried to evolve a scientific method for computation of life valuation and compensation for death or total and partial disablement on the basis of general laws of the land and other relevant laws. Considering the average age of the workers and the expected age of the spouse (and dependant), the Manch counted the average income per head (GDP, the Gross

Domestic Product per capita), inflation rate (change of consumer price index for industrial workers over the previous year), interest rate (maximum interest on deposit), value of Re 1 after one year, by making adjustment for inflation (@ 10 per cent compound per annum at an interest rate of 12 per cent compound) and the sum of present values of future amounts of Rs 12,000 per annum adjusted for future inflation (@ 10 per cent per annum compound) at an interest rate of 12 per cent per annum compounded for 35 years. This came to more than Rs 3,00,000. The Manch suggested a second method of computation, taking into account the minimum rate of wages in stone crushing units, average age of victims both dead and alive, average number of years of service left, taking 60 years as age of retirement and a gross loss of earning. On this basis, the compensation amount came to Rs 2,67,000. But strangely, the Labour Commissioner suggested that the dead workers were entitled to compensation of Rs 50,000 each and the surviving suffering workers should be compensated in accordance with the principles laid down under the Workmen's Compensation Act.

The court, in its final order of 26 November 1996, found the suggested compensation for the dead workers 'on the lower side'. But without going into the issue of determining the principles and procedures for compensation in the absence of clear-cut rules, it awarded an amount of Rs 1,00,000 each. For the rest of the suffering workers, the court directed for the formulation of a compensation formula under the Act. The court also directed that in case the compensation amount was not paid by the factory owner, it should attract interest at 12 per cent per annum.

In Chinchurgheria village, where Surendra Khanij was situated, there is not a single household which has not seen premature and painful death. It is a village of twenty-five households, who have nothing else to live on except wage labour. They go to the nearby markets and towns for manual work.

The first death among the workers of Surendra Khanij was reported in January 1991, but soon it became a regular happening. By the end of 1992, at least ten more workers died, suffering from fever, cough and breathlessness. The death count continues even today. Sixty-five-year-old Matal Hansda has lost his daughter and son who were working at the factory, and his only surviving son is also a victim of chronic silicosis. Dukhu Hansda, Ram Hansda and Dala Hansda were three members of the same family who died. Now the only adult person alive in the household is Sukhul Muni Hansda, wife of Dukhu, with three small sons to care for. Sita Murmu lost both her husband and brother-in-law. Kanai Soren has left his widow Chura Muni Soren with two sons and two daughters. She works as an agricultural labourer, but gets work only for 4–5 months in a year. Weeping intensely, she reported how she forced her children to drink some mahuva, to douse the pangs of hunger. Kanai Murmu was one of the first workers of Surendra Khanij to have died in 1991. His widow Baso Murmu works in the local market and takes care of her two sons and two daughters. Her 10-year-old son also works in the market.

Pasro is another tribal village which supplied workers to Surendra Khanij. It has more than a hundred houses of tribal Santhals. Landless, they work in factories, markets, or various government schemes. Sunil, a former worker of Surendra Khanij, was waiting death in this village. Reduced to a mere skeleton, he pleaded for some monetary help from whoever visited his place. Unable to breathe, he cried to express himself. A deep sense of helplessness pervades the village, which has nothing to look forward to except disease and death.

GUILTY INDUSTRIES

Radharaman, working as a reeler in the reeling department in a cotton mill in the Taratala area of Calcutta since 1968, started suffering from acute respiratory problems since 1989. Severe chest congestion and cough became his constant companions. In January 1991, he went for a check-up at the Employers' State Insurance hospital, where 'byssinosis' was mentioned in his prescription but with a question mark. Admitted to an ESI hospital for more than four months in December 1992, he was finally diagnosed for cotton induced bronchital asthma (byssinosis). The doctor strongly advised him to avoid cotton dust at his work place, but after discharge from the hospital, he was forced to work in the same department. In March 1993 his condition again became critical. At ESI hospital, he was advised to go in for a lung function test, for which he had to pay a private doctor. In July 1993, he applied to the Regional Director, ESI, to allow him to appear before the Special Medical Board. In March 1994, he was produced before the Medical Board, which decided on a 40 per cent disability for him and gave a compensation of Rs 15.68 per day since December 1992, when he was first admitted to the ESI hospital.

At this time, the factory disallowed him to rejoin work and refused his case of redeployment. The issue finally reached the Labour

Commissioner, and after several dates and discussions at a tripartite level, he was redeployed in gas welding as a helper in October 1996. In all these years of his pain, he said, nobody except a social organization, came to his rescue. The unions and the Labour Commissioner were all asking him to take some lumpsum payment from the management and leave the job. There were many more workers suffering like him in the mill, but scared of losing their jobs, they held their silence. He had prepared a list of 47 fellow workers in a plight similar to his.

Shakti Dutta, 48 year old, worked in Waldies Factory, Kann Nagar in Hooghly district, making lead dust in the paint factory since 1970. Suffering form lead neuropathy, he could not move one of his hands since the last two years. When he went to the ESI in early 1995, he said he was in deep pain, able neither to work nor to sleep, but the ESI did not admit him. He was asked to get all the tests done outside. After a lot of problems, he was admitted in an ESI hospital in May 1996. He is still unable to do much work with his disabled hand. Shakti was also refused redeployment by the management and was forced to continue with the same job even during his treatment. Only in March 1997 the management agreed for his redeployment on the persuasion of ESI director, Inspector of Factories, Labour Commissioner and others. Samir Kumar Rai, another worker in the same factory since 1970, also suffers from lead neuropathy, with very little grip in his hands. Recently, on the intervention of the Nagrik Manch, the Medical Inspector (Factories) ordered blood tests of all the workers of the factory. In the meantime, the management floated a voluntary retirement scheme for the workers, indicating its plan to get rid of the ailing workers.

Sanat Kumar Paul, being a casual worker in Eco Battery Industries, Belghachia Road, Calcutta, was not covered under the ESI. He started having problems with his hands in 1995. The private and ESI treatment confirmed his disease as lead neuropathy, but the management retrenched him in the same year. The bipartite and tripartite meetings held under the Labour Department yielded no results.

FUMBLING FOR SOLUTIONS

Everybody in the official circles in Calcutta agreed on one thing: the Supreme Court judgement on workers and villagers of Chinchurgheria was a landmark in terms of protecting the workers' lives in a hazardous environment. But they also gave several reasons why they could not implement even its basic elements, why the judgement could not be followed up in the state, or why it had not been possible for the state machinery to initiate action against the culprit industries.

Mr B. Mukherjee of the West Bengal Pollution Control Board said that the WBPCB had no role in the matter and neither had the Supreme Court asked it to monitor the implementation of the judgement in any manner. There was, as such, no provision in the Environment Protection Act to deal with the sufferings of workers. Subesh K. Das, Labour Commissioner, said that this type of judgement created a precedent, but it was also necessary to consider whether it was possible to implement such judgements. The employer's capacity to pay should be taken note of. In the informal sector, it was difficult to pay one lakh rupees all at once.

Apparently, a notice was issued to the management by the Labour Commissioner for collection of the compensation money but the management said that it was not in a position to pay. Then the owners advertised for sale of company property, but there were no buyers. Next, the Commissioner filed a case against the management under the Workmen's Compensation Act as a last resort. For the ailing workers, he is pursuing investigation as per the directions of the court.

Though admittedly there were many problems of occupational health and safety in the state, he said that it was not possible to inspect each and every factory in his jurisdiction. There are only two Factory Inspectors for the 2000 registered factories in Howrah district and there are twelve vacancies out of the eighteen sanctioned posts of Factory Inspector (Medical) in the entire state.

V. Subramanian, Labour Secretary, did not see any need to establish a parallel health system on occupational health, when state hospitals were taking care of workers' health, 'and they are doing a fine job'. The issue of occupational health was very wide, he said. Doctors, engineers, air hostesses, journalists, everybody suffered from occupational diseases. One could not take care of the health of every citizen in the state. He did not wish to be seen as defending environmental hazards as such, but it was better to work for twenty years and die of occupational diseases, than die of hunger, owing to closure of a factory in the name of pollution and hazards. It was also not possible to detect occupational diseases of workers, especially in the informal sector. More than governmental action, it was awareness, education and training among workers and management that was needed on this front, he said.

In spite of this lacklustre response, the government does seem to be doing something for the first time and has recognized the need to launch some concrete programmes. Shanti Ghatak, the State Labour Minister, stated that in 1997–8 itself, they had opened an occupational disease centre at the ESI Hospital, Thakur Pukur. At Manik Tola ESI Hospital, there was a medical board, constituted for workers suffering from occupational diseases. At Belur ESI Hospital, an occupational disease detection and treatment centre was opened. Now there were at least two ESI hospitals in West Bengal which could detect and treat occupational diseases of workers covered under ESI. Since 1996, thirteen cases of occupational diseases were reportedly detected at the ESI Hospital, Belur.

On 5 and 6 April 1997, a workshop on occupational diseases and hazardous environment was organized at Belur, Howrah, by Shramjivi Hospital, Nagrik Manch, and Bank of Baroda Employees Union. Participating also were victims of occupational diseases, representatives of some central trade unions, Regional Labour Institute, Inspector of Factories, and ESI Hospital, Belur. Naba Dutta, General Secretary, Nagrik Manch, while commenting on the initiative, said that an impression was gaining ground that all environmental concerns and judicial activism based on it, were against their employment, since it was leading to closure of factories on the grounds of pollution. The Nagrik Manch therefore wanted to redefine the issue of industrial pollution and environment protection as also an issue of workers' health, of their life and longevity, of their physical and mental well-being. It wanted to establish the responsibility of the employer not only towards the environment but also towards his worker.

Under the relevant acts, he pointed out, there were provisions of jailing, fines up to Rs 5 lakh, attaching of property and so on. But this was not happening. Only simple closure was taking place, and it was the worker and not the owner who was getting punished. The closures were not taking place under the Industrial Disputes Act, which could have given the people some scope of protest, but under Supreme Court directives. In these circumstances, a new slogan was needed as well as a new approach to address the situation.

A leading activist of this initiative, Ashim Roy, viewed the situation through a different perspective. According to him, the initiative, from a citizens' group like his, needed to start from above initially. It could then be transformed to activities from below. His was a three-tier aim: to establish some infrastructure, to disseminate some information and, based on that, build up some cases of compensation, leading to a larger movement.

What had been the gains, other than the Supreme Court judgement? Naba Datta replied that they had started from scratch two years earlier. Now they were a few steps ahead, but even today, it had broadly remained a legal battle. Even within the legal, administrative confines, many tasks remained unfulfilled. First, occupational diseases had to be detected and diagnosed under the government machinery, but this was inadequate in itself. Second, there had to be preventive measures, including redeployment after discovery. This was non-existent. Third, there had to be proper compensation. There were some damages, like lung damage, which were calculable, but the damages on skin, wrist etc. were not easy to calculate. Norms and guidelines to assess these damages needed to be evolved.

This particular initiative and its role in raising consciousness on the issue of occupational diseases had also galvanized various central trade unions like AITUC, HMS, UTUS, UTUS(LS), and INTUC which nevertheless accepted their shortcoming. They now realized the need to integrate the issue of health and safety in their programmes, especially in the context of heightened environmental consciousness and activism. It was also clear during our interactions that they were yet to formulate any concrete programmes.

Ranjit Guha, Secretary AITUC, acknowledged that trade unions in the state were negligent on this issue, which had aggravated this situation. Even in the public sector as in the Calcutta Port Trust, the reality was alarming. Working conditions in a large number of chemical factories in and around Calcutta were dangerous for health. The government was shifting the tanneries, but no measures were taken to improve their technology and working conditions. Trade unions accepted this shifting without questioning.

According to Samar Chakravorty, General Secretary, INTUC, the Supreme Court judgement had opened the floodgates. It needed to be taken in its wider dimension. The judgement was not only about compensation, money etc., but also about the quality of man and living, and about quality production and productivity. If a worker was not a quality man, he could not produce quality goods. The question of the physical formation of a worker and the humanization of the work place needed to be an important agenda for all of them. If a worker became sick or a victim of pollution hazards in a competitive environment, he would become a social parasite. The present attitude of the employer as well as the government was socially unacceptable and economically unproductive, he said. Prior to 1993, not a single worker in the state had been given compensation on account of occupational diseases under the ESI. Trade unions must now give up their dogmatic ideas and accept the commendable role of support groups, social organizations, and NGOs in this field.

Ashok Ghosh, General Secretary, UTUC, pointed out that occupational health was a problem. Though not as valid as the problem of employment, the government could not ignore it under any pretext. The trade unions should have put pressure on the government to activate the machinery which they had created.

A Beacon: The Shramjivi Hospital

Shramjivi Hospital, 3.5 kilometres from Howrah Station and a walking distance from Belur Math, is an institution founded by workers of a closed industry. Started to meet the needs of ailing workers, it has now become an essential part of the community, with the local teachers, artists, doctors, social workers, and students becoming involved in it in one way or another. Surrounded by factories, mostly closed, small shops and houses, Shramjivi Hospital exudes a homely ambience. The building is an old house of five or six rooms, a staircase on its veranda and a first floor with a tin-shaded working space. The rooms are spick and span, and filled with patients and their attendants. Some nurses dressed in white bustle about, but the local workers normally do several important and regular works of cleaning, helping, etc. They make the iron beds, stretchers, lights, tables, benches, and so on themselves.

Twelve doctors come in turns twice or thrice a week, and give specialized services to patients. Every evening there is a general outdoor clinic free of cost. The eye or skin specialist, the paediatrician or gynaecologist charge only Rs 11. Round the clock, there is at least one doctor in attendance. Fifteen trained volunteers come in shifts, ensuring that two of them are always present in the hospital. Other volunteer workers are also available to support the regular staff.

Chitranjan Paul, 53, is one of the volunteers. He worked for Indo–Japan Steel Ltd. After the closure of the factory, Chitranjan started a petty shop. In July 1996, his gastric ulcer became serious. He was first taken to the outdoor clinic after vomiting blood. The doctors advised him to shift to some other hospital, but he was adamant on being treated by them. Having recovered, he spends his entire day in voluntary service at the hospital.

Complex functions like MacMurry osteotomy, eye lens transplantation, plastic surgery, external fixation in different types of external fractures, etc. are done here. The charges for an operation and post-operation care for hernia are Rs 975, compared to Rs 5000 elsewhere. For appendicitis, the maximum charge is Rs 1500, compared to 7–8 thousand elsewhere. Ordinary blood test is charged Rs 7, blood-sugar test Rs 11, and ECG Rs 21. The hospital has carried out more than 350 operations.

Shramjivi took shape because there were trade unions like Indo–Japan Steel Ltd Employees Union, which wanted to take up the issue of workers' health in a constructive manner, and also to do something for society at large. There were associations of doctors, like the People's Health Service Association, which wanted to work among the working people. Both came together and started this modest initiative in 1983. At first, the hospital was at a very small and distant place, with only outdoor facilities. It was shifted to a more convenient place near the National Iron and Steel Factory in 1989, and some investigations, x-ray and minor operations were started. In 1993, the Indo–Japan Employees Union acquired the present premises. The union purchased more than a tonne of iron scrap and the workers, after their working shifts in the factory, fabricated operation tables, lamps, trolleys, chairs, etc. out of the scrap. The in-house facilities commenced with four beds and an operation theatre. The finances came from the workers' voluntary contributions of Rs 10 per month. The hospital has never asked for or accepted contributions from any corporate groups or foreign agencies. But ever since Indo–Japan Steel closed in July 1996, the hospital's future has become a question mark. More than a hundred local people have decided to regularly contribute to the hospital. They organized a Save Shramjivi Convention in March 1997 and took out a demonstration for mass collection. Some artists also performed in 1996 to collect funds. They have decided to make this an annual feature.

Against all odds, the hospital decided in April 1997, to set up a bi-weekly Occupational Disease Detection Centre, which is rare in West Bengal. To deal with industrial accidents, they would also like to develop facilities for micro vascular surgery.

In Someone Else's Backyard

The residents of Rajasthan, Haryana, and Uttar Pradesh, the states adjoining the national capital city are angry against the relocation of several polluting industries on their agricultural land. Mubin Khan, from a village in Bhiwadi district in Rajasthan expressed his discontent: 'Our village has more than 700 bighas of agricultural land, all of which comes under the green belt. We do not want our agricultural areas to be transformed into industrial ones. It is not employment generation, but destruction.' Another complained that Delhi's search for clean air will spread pollution to areas that were clean before.

Polluting industries in Delhi, which have been affected by the Supreme Court's order, are being invited by the Rajasthan government to its industrial estates in Bhiwadi, some 70 kilometres from Delhi, Neemrana (110 kilometres) and Alwar (150 kilometres). Additional chunks of land were acquired by the Rajasthan State Industrial Development and Investment Corporation (RIICO) in 1997, for expansion of the Bhiwadi industrial area.

The Bhiwadi industrial area was set up in the mid-1980s on 200 acres of acquired land. The industrial estate is now spread over 2000 acres and threatening to devour more land in the region. Says an official of the RIICO: 'An area of 10,000 acres of land in Alwar district will soon be acquired. The state government is even providing capital investment subsidy and exemption from sales tax as incentives.' An official spokesperson from the National Capital Region Planning Board confirmed the government's plans to acquire land on a massive scale. Industrial estates developed in Dharuhera (Haryana) on about 670 acres for medium-scale industries and adjoining Bhiwadi (Rajasthan) on about 1,160 acres for small-scale units can accommodate polluting industries from Delhi, he said. Active consultations

Published in *Inter Press Service*, 12 September 1998.

with the Rajasthan authorities are going on. In the UP sub-region 4,901.95 acres and in the Rajasthan sub-region 5,303.61 acres (Alwar, Bhiwadi, Neemrana, Behror) have already been identified. The Rajasthan government is taking steps to acquire an additional 3,095 acres. It has also offered acquisition of land outside its industrial areas for industries which can not be located in the regular industrial areas. These plots could be allotted to industry. If industries by themselves procured such land, change of land use could be granted.

A high-level team of officials was constituted in Uttar Pradesh to smooth the way for relocating polluting industries. The Delhi government said it has started a single-window service to facilitate the evacuation process. Authorities in Delhi were meeting their counterparts in the neighbouring states every month. At least one meeting of the single-window service was held every fortnight in Delhi.

It seemed the villagers had no choice but to give in. The district collector had apparently issued orders directing the villagers to sell their lands. Then the tehsildar, the naib tehsildar, the sub-registrar and the patwari camped in a house in the area. As the armed police kept vigil, the villagers were taken to the house and their thumb impressions were taken on sale deeds, alleged an aggrieved resident of Sare Khurd village in Alwar district. Pritam Singh, another villager was threatened with murder when, unwilling to sell his land, he sought an unrealistic (then) Rs 2 lakh per bigha as minimum compensation. The standing crop on another plot of land was destroyed and he was beaten up. While Pritam Singh was still holding on to his land, the Kedias, who had purchased the surrounding plots, fenced off the area, making it inaccessible to him. Jungroo Bai's husband, Ganga Singh was escorted to the place of registration by a middleman. When he refused to sign the papers, he was beaten up allegedly by the Kedia men and his thumb impression was taken forcibly.

Mangal Singh alleged that the hirelings of the Winsome Breweries, threw out his belongings from his land. When Mangal Singh arrived on hearing the news, the area SDM told him that the land had been acquired by the RIICO. He could only express his surprise as he had received no notice in this regard. All his efforts to contact officials of Winsome Breweries proved futile. There was heavy police deployment at the gates of the company office when we visited the place.

This was not the first time that the polluting industries were relocated outside Delhi. The Supreme Court in 1992 decided that stone crushing plants were a health hazard and ordered that they should move from along the capital's border to Haryana. Following the court order, the Haryana administration acquired 2025 hectares of land belonging to two villages on its side of the border where it relocated the stone crushing plants that supply gravel to Delhi's booming building industry. Not only had the village women and herders lost grazing grounds and forests where they picked firewood, but the village of Pali was now shrouded in the fine dust spewing from the primitive plants. Everyone complained of respiratory diseases here. Angry with the authorities, a villager Ram Dayal asked: 'Are our children immune to tuberculosis? Are the lives of our children less precious than those of the city folk?' Another villager was equally bitter: 'If Delhi needs gravel for its buildings, it should also be prepared to suffer the evils of pollution or pay for improving the conditions. Why should it simply relocate its problems in another's backyard?' he asked.

Caught between Borders

Damaged Indian boat, Rameswaram

A mother's loss, Pamban Village, Tamil Nadu

❦

The Case of Indian and Pakistani Fishermen
The Other Border Question

humari jaat machimaar
humari naat machimaar
hum sab machimaar ek
(Our caste is fishing, our occupation is fishing,
we fishermen are all one.)

This sentiment of the fisherfolk of Varanwada village, in the Union Territory of Diu in western India would be echoed by the fishermen from Pakistan's Sindh province further up the Arabian Sea coast. But governments are no respecters of sentiment. A large number of Indian and Pakistani fishermen are lodged in jails in each other's country, deprived of basic legal and human rights, arrested for transgressing the maritime boundary between the two countries while engaged in fishing.

Following a bilateral agreement, each side released a batch of 194 fishermen on 15 July 1997. The exact number of those still imprisoned in the two countries is not officially known. The Fishermen's Cooperative Society in Karachi has stated that around 130 Indian fishermen remain in Pakistani jails, while the corresponding figure for Pakistani fishermen in Indian jails is estimated at 118. Consistent efforts by the agencies concerned have made a limited exchange of fishermen possible in recent years. However, their task is made difficult by the 'exchange protocol', being somewhat similar to the procedure followed in the case of prisoners of war. Even after completing their term of punishment as per court orders, the fishermen are not

Published in *Himal South Asia*, December 1997.

released. They have to wait for years for a formal process of exchange of prisoners to take place.

Nevertheless, for the first time, trade unions and labour support groups of both India and Pakistan, and their common platform, the South Asian Labour Forum (SALF), was successful in drawing the attention of the authorities concerned to the plight of the fishermen. The SALF had demanded the unconditional release of all the detained persons and a stop to the mid-sea arrest and imprisonment of fishermen. The SALF's Indian chapter, which consisted of central trade unions like AITUC, CITU, HMS, and HMKP and labour support groups like CEC, welcomed the exchange of prisoners, but criticized the absence of a clear policy of action to prevent the arrest and detention of innocent fishermen. H. Mahadevan of AITUC said that the act of exchange currently was devoid of any clearly worked out policy, and was done in the absence of a bilateral agreement defining the maritime borders between the countries. This did not give any guarantee that arbitrary arrests and illegal detention would not happen again. The Forum demanded a bilateral agreement between India and Pakistan that clearly defined the maritime boundaries; the establishment of effective steps to make the boundary visible to the fishermen in the sea; and a regional agreement at the SAARC level, whereby the fishermen of South Asian countries could fearlessly and in a friendly manner, fish in the Arabian Sea, the Indian Ocean, and the Bay of Bengal.

India, Pakistan, Myanmar, Bangladesh, Sri Lanka, and the Maldives share the water and the resources of the Bay of Bengal, the Indian Ocean and the Arabian Sea. India itself has a long coastline of 7417 kilometres. Pakistan's coastline is adjacent to that of the 1640 kilometres long Gujarat coast of India. In the Bay of Bengal, India shares an adjacent coastline and thus a part of the Bay's marine resources with Bangladesh, Sri Lanka has its northern coastline along the Palk Strait. So far, however, there are no bilateral agreements on maritime boundaries between India and its South Asian neighbours. Also, the Maritime Zones of India Act, 1976 and 1981, under which the fishermen are detained and punished, do not correspond with the UN Convention of the Law of the Sea, of which India is a signatory. The same is true of the Maritime Zone of Pakistan Act, which is virtually identical to that of India. For fishermen, who venture out on the seas to eke out a living, the concept of marine borders is difficult to comprehend.

Pakistani Fishermen in Indian Jails

Thirty-one-year-old Ghani Rehman has been in the Porbander jail of India for two and a half years. Ghani was the captain of Al-Jashan boat which had on board fourteen fishermen. All were caught and penalized. The rest were punished for two years and two months, and after having completed their period, they were moved to the police headquarters on 1 January 1997.

Residing in the Sarhad state, Ghani's father Sayid is a fisherman as are his three brothers. Their fishing expedition to the ocean lasts 10–15 days. Hoping for a better income, Ghani went to Karachi and started working on the boat of a businessman. Ghani narrated the events: 'This time we were on the ocean for more than a month. We were throwing our net to catch fish and it was impossible to make out where the wind and water were driving the boat. If there were some signposts on the ocean, it would make it easier to discern boundaries. Suddenly the navy came. We did not even know whose navy it was. And then we were all captured in the Indian check-post.'

After being arrested, Ghani wrote a letter to the boat owner, but no reply came. He wrote to his family to do something but they were helpless. Ghani's wife wrote to him but did not talk of her difficulties. But her helplessness can be imagined.

Muhammad Alam, 18, a seaman of Al-Kabutar boat has his mother, father, wife, and two children in Karachi to look after. On the morning of 8 January 1988 he boarded the boat with five others for fishing. At night they cast the net and were half asleep. Dawn was just breaking when the Indian coast guard caught them. The boat was kept at Porbander. They were sent to the police station and then to the Porbander jail.

Naushad Ali with eight others was on the Samira boat when they were caught on Indian waters on 8 October 1989. Going through the tortuous process of captivity, police custody, and court cases and after completing their interlude, they were sent to police custody at Kutch in September 1991. Since then, they were in police custody. Waiting endlessly for his release and return to his country, Naushad Ali was completely shattered. He asked forlornly, 'Why should we bear this pain because of tensions between the two powers? Our heart is dead. Our hopes have been constantly belied.' Sikander and six other members of his family from the Sindh state, were caught while fishing, in 1994. They did not know when they had entered the Indian border in the ocean. Even after completing their jail term, they kept asking what their crime was.

INDIAN FISHERMEN IN PAKISTANI JAILS

The two villages of Varanwada and Saudwara, in the Union Territory of Diu, were a catchment area for fisherfolk, not only of these two villages but also of the nearby areas. There were more than two thousand fisherfolk in these two villages and almost the same number had come from outside. All used to go to far-off regions in the ocean to catch fish. Both villages were full of narratives of anguish of fishermen caught by the Pakistani coast guard or navy on the charge of trespassing. According to Lakhan Bhai Puja, chief of the Boat Association Vari Vistaar at Varanwada, in the past twenty years, more than five hundred fishermen of this area had been arrested by the Pakistani coast guard and more than 150 boats seized. One fisherman of Nawa Bandar died in a Pakistani jail.

Dhanji Harji Rathod wrote to his family from the Landeo Jail, Barrack No. 11, Karachi East 34, 'Jail was our destiny ... . They caught us by force in the ocean. For five days we were kept in the boat itself. Then they took us to the jail. We got one cup of pulses and two slices of bread to eat. The bread was baked only on one side. We do not wish this punishment even on our enemies. There was one Pakistani prisoner who helped us sometimes and gave us cigarettes and soap.'

THE SCARED CHILDREN

Fifty children in these villages, who had gone with their fathers on the boat, were caught and kept in Pakistan jails. For years they stayed in jails or the Idhi Centre. These children had lost the innocence and laughter of childhood.

Some months back, the Pakistan government had released 38 such children. A now-free eighteen-year-old Manji Dayar of Varanwada village of Diu fearfully remembered that day of 1994, when he was caught in the ocean. It was early morning when suddenly there was firing in the air. The Pakistani navy stopped the boat and cut the net. The five persons on the boat were caught and taken to Karachi. There they were kept in police custody for three days and then jail, then Idhi Centre. On returning home, Manji had become a wage labourer. He did not venture to the ocean at all.

Manji did not want to remember his old life, but he was very bitter about the present. When he and the others returned, there was a meeting at Diu. The collector, the commissioner, all came. All promised help. They even filled a form, but it all came to nothing. If the son of collector or the commissioner, he said, had gone to jail, they would have realized the pain it causes. Manji was worried for his family and about the future. He did not want his brother to be obliterated like him. He wanted him to get educated.

Though twelve-year-old Nanji Murji was released from jail, his father continued to be in the Karachi jail. Nanji had accompanied his father for the first time during a school holiday. Nanji stayed in the jail for five months and then was sent to Idhi Centre. Now he did not want to go to school.

The territorial waters of India are controlled by Coast Guard, BSF, Customs and Army Inland Units. According to K.C. Pande, Commandant, District Headquarters, Coast Guard, at Porbander, there are no signals on the sea to demarcate the sea border. Above all, there is no agreed boundary on the Arabian Sea between India and Pakistan. For mutual convenience, the patrolling agencies have worked out an imaginary line along the Sir Creek region, off the coast of Kutch. The fishing boats could unwittingly cross into the other country's territory because of tidal currents, wind force, cyclone, and engine failures. The captured Pakistani fishing boats had no navigational aids. Also, no Pakistani fishing boats were found with arms and ammunition on board. But this does not stop the Coast Guard from continuing to arrest the fisherfolk. The Coast Guard officials also candidly admitted the practice of 'tit for tat' among the enforcement agencies patrolling the territorial waters where 'they capture so many of our boats and we capture that many in retaliation'.

The superintendents in Porbander jail clarified that they do not inform the Pakistani High Commission on their own. Consular access for the Pakistani prisoners is available only in the Central Jail of Jodhpur, Rajasthan. But they are sent to Central Jail, Jodhpur for consular access only on the orders of the Union Home Ministry, routed through the Gujarat government.

There are a large number of fishworkers who have completed their period of conviction, but have to wait in police camps at Porbander for the exchange to take place. The so-called Police Headquarters at Porbander become a new jail for them. Atul Karwal, DSP, Porbander, who was in-charge of these particular cases of 'freed but fettered' fishworkers, accepted that these people should have been deported immediately after the completion of conviction.

The government officials also had some suggestions to offer, based on their specific experiences. The DCP suggested that there should be a permanent structure to house the detained fishworkers who had completed their conviction. The Coast Guard Commandant suggested that there should be visible demarcation lines on the sea. In case of violations, the boats needed to be seized, the crew and fishermen fined and released. Further, there had to be separate and speedy courts for this type of offences.

R. Venkataramani, senior advocate, Supreme Court, has done a study on the legal aspect of the problem, on behalf of the Indian chapter of SALF. In case of violation of territorial waters or exclusive economic zones, he said, the intruding fishermen were booked under the Maritime Zones Acts, 1975 and 1976 by the Pakistani Maritime Security Agencies, and under the Maritime Zones Acts, 1976 and 1981 by their Indian counterparts. Violation of both the Acts was punishable by imprisonment and/or imposition of a hefty fine. Section 10 of the MZI Act 1981 of India says (a) ... where such contravention takes place in any area within the territorial waters of India, be punishable with imprisonment for a term not exceeding three years or with fine not exceeding Rs fifteen lakhs or with both; and (b) ... where such contravention takes place in any area within the exclusive economic zone of India, be punishable with fine not exceeding Rs ten lakhs.

He clarified, however, that Part 2 and 3 of article 73 of the United Nations Convention on the Law of the Sea, 1983, stated that arrested vessels and their crews shall be promptly released upon the posting of reasonable bond or other security. Coastal state penalties for violations of fisheries laws and regulations in the exclusive economic zones may not include imprisonment, in the absence of agreements to the contrary by the states concerned, or any other form of corporal punishment. It was clear, he said, that the MZI Act of 1981 violated the spirit of the United Nations Convention and did not fully adhere to the International Law of the Sea. The worst human tragedy, he said, was the completely illegal and wrongful detention of persons who had already completed their jail terms. Their detention was clearly in contravention of article 21 of the Indian Constitution by which guarantees were available to all persons, including aliens and foreigners. An important point that had to be considered was whether the fishermen had crossed the maritime boundary intentionally. All the evidence and documents showed that they had not, he pointed out.

For many years, the fishworkers' unions, boatowners associations and trade unions of both countries have been asking their respective governments to work out a long-term solution. Since 1988, Shree Akhil Gujarat Machhimar Mahamandal, Fishermen Boat Association, Diu, Porbander Machhimar Boat Association, Gujarat Marine Products Exporters Association, National Fishworkers Forum and others had sent several petitions and representations to the Indian government. When labour leaders from India, Pakistan, and Sri Lanka, representing the SALF, met the then External Affairs Minister, I.K. Gujral and the Home Minister, Indrajit Gupta on 4 December 1996, drawing their attention to the problem, both assured immediate and long-term action. Gujral confirmed that no charges were levelled against any of the detained fishermen except their violation of territorial waters.

The Fishermen's Cooperative Society Ltd., Karachi, the Fishermen's Union, Karachi, Pakistan Institute of Labour Education and Research, and Pakistan Chapter of SALF, had taken up this issue with the Pakistan government. Karamat Ali, convenor of the Pakistan chapter of SALF, in a letter to Mohammad Nawaz Sharif, Prime Minister of Pakistan, commented that as long as the two governments could not take decisive action even in a matter that did not have anything to do with issues of the national security of either country, but was surely a question of continued violation of the fundamental human rights of these fishermen, there was no point in claiming that the two governments were sincerely working for the improvement of mutual relations. A team went to all the jails in Sindh where Indian fishermen were detained, and collected information about them. They also met 242 Indian fishermen detained there and delivered to each of them a packet containing blankets, coats, towels, chappals, medicines, and other basic things for daily use.

Fisherfolk continued to be detained on the high seas. While the authorities of the two countries invariably cited the issue of national interests, they appeared to have ignored two major questions involved—the fisherfolk's right to resources and livelihood and the incompatibility of their national laws with regard to the seas and internal laws and conventions.

The Killing of Indian
and Sri Lankan Fishermen
Tomorrow is Another Country

Pathinathan, S.P. Royappan, Susha Raj, John, Sebastian, M. Sahayam, and Pandi were among the thousands of marine fishers of Palk Bay, in the Tamil Nadu state of India, who fished in the seas bordering India and Sri Lanka. They got arrested, injured, harassed and even killed in the sea. Their boats were drowned or captured, never to be returned. K.S. Nicholas, W. Wilbert, K.S. Joseph Washingtoo, Sirinimal Fernando, Wijendra Waduge Chandra and many more fishermen of Sri Lanka met the same fate in the seas at the hands of the Indian navy and coast guard. Both were being charged for intruding into the sea territory of the neighbouring country. With the continued problem of terrorism in India and ethnic strife in Sri Lanka, they were now even dubbed as terrorists or militants in the seas, and treated as such.

Sir Lanka and India share a maritime border more than 400 kilometres long. In the Palk Bay region, where the Kachchativu island is located and which divides the coastal regions of Nagapattinam, Quaid-e-Millath, Thanjavur, Pudukkottai, and Ramanathapuram districts from the Republic of Sri Lanka, firing, detention, and jail of fishermen is an almost regular happening. On 22 February 1998, an Indian fisherman, Krishnan, of Menandai near Rameswaram, died at sea on his boat, when the Sri Lankan navy opened fire on them. On 4 March, the Indian coast guard arrested three Sri Lankan trawlers and their fifteen crew members in the Gulf of Mannar on charges of violating the Maritime Zones Act. On 11 May, an Indian fishing boat was fired upon off Talaimannar and out of a five-member crew, three were arrested.

Published in *Himal South Asia*, November 1998.

In 1997, on 3 June, five Sri Lankan fishermen were arrested in Indian waters. On 10 June, Ganesan and three other Indian fishermen were fired upon by a Sri Lankan patrol boat. Ganesan was injured on his legs and shoulders, and other fishermen escaped into the sea. On 12 June, four fishermen of Rameswaram were fired up by a Sri Lankan naval vessel. One fisherman, Lourduraj, fell inside the boat with bullet wounds and the other three jumped into the sea and swam to safety. On 18 July, two Sri Lankan naval helicopters allegedly sprayed bullets on a motorized country boat off Nagapattinam coast, killing two fishermen on the spot. In 1996, the then External Affairs Minister, I.K. Gujral informed the Indian Parliament that eleven Indian fishermen were killed in twenty incidents of shooting at Indian fishing vessels. The Sri Lankan navy acknowledged its involvement in three incidents, but denied responsibility in the other reported cases.

Since 1983, till the end of August 1991, the Tamil Nadu government reported, 'there have been 236 incidents of attacks by the Sri Lankan navy on Tamil Nadu fishermen. 303 boats have been attacked and 486 fishermen have been affected. 51 boats have been destroyed. 135 fishermen have been attacked and injured. Over 50 fishermen have been killed. 57 fishermen have been injured in firing incidents. The Sri Lankan navy has seized 65 motor boats and arrested 205 fishermen. There has been an increase in these incidents, particularly this year (1991).' In November 1993 the Tamil Nadu Government reported that in the past three years, 25 Tamil Nadu fishermen had been killed in firing by the Sri Lankan navy. Besides, 109 were injured. A total of 136 attacks had taken place on fishing boats; 15 of these had sunk.

THE SUFFERING INDIAN FISHERMEN AND THEIR FAMILIES

Rameswaram is an island in the Ramanathapuram district, situated in the south-central Bay of Bengal coast of Tamil Nadu. It has an approximate coastline of 80 kilometres, a land area of 38 square kilometres and a fishing community of around 35,000, in which the active sea-going fishermen are around 8000. To the island's south is the Gulf of Mannar and to the north, Palk Bay. On the eastern tip of the island is Dhanushkodi from where it is only about 16 kilometres to Talaimannar of Sri Lanka.

Saghai Nagar, a settlement of a thousand fishermen in Pamban village near Rameswaram, has been a witness to firing, injuries, detention, and damaging of boats in the seas by the Sri Lankan navy. On the night of 5 November 1996, Pathinathan and three others from the settlement went for fishing. It was a stormy night, and their boat crossed to the Sri Lankan side near Talaimannar. When they realized this, they turned back, but suddenly they were confronted by a boat of the Sri Lankan navy. The fishermen put on the light, raised their hands and begged for mercy. But the men in the naval boat kept firing and shouting at them. They were captured, their shirts taken off, blind folded, beaten up, and lodged in the Mannar jail. Next day they were produced in a court in Kothi Manar. They were in Mannar jail for two more days and from there were taken to Marihana police station for a week. After ten days of their captivity, somebody from the Indian Embassy visited them and assured them of justice. But they had to spend a hundred days in several police stations and jails before they were taken to Jaffna and handed over to an Indian naval ship in mid-sea. They had to go to Talaimannar in Sri Lanka on five occasions to get the boat back. When after four months of their release they got the boat, it was badly damaged and without the engine. Pathinathan is no longer a boat owner but only a boat worker heavily indebted.

Susha Raj of the same settlement took a loan from a bank to purchase a boat worth Rs 5 lakh. He and six others went to fish from Pamban in the late evening of 5 January 1995. At night around 11 o'clock, near Kachchativu, a Sri Lanka naval boat came and started firing at them indiscriminately. Susha Raj alone was left alive. The bodies of his companions were later found in the sea. Susha Raj is now a coolie with heavy debt.

John and four other fishermen were all in a mechanized fishing boat in the late night of 14 July 1997, when they were fired upon by a Sri Lankan naval boat. The boat was badly damaged and drowned. Two men died on the spot. The rest swam to an island called Nedund-hevve on the Sri Lankan side. There some Sri Lankan fishermen rescued them and a local organization took them to Jaffna for medical treatment. After hospitalization for five days, they were jailed in Jaffna and produced before a court in seven days. There were no cases against them, but after every fourteen days, they were taken to the court and then sent back to jail. They were in Jaffna jail for five and a half months. Leave alone compensation, the families of the dead fishermen had not even got the death certificates. Sebastian's right hand is permanently crippled.

Xavier, another fisherman of the same village, maintained that the government's figures of deaths and injuries were only partial. In the past seven years, he said, around six hundred Indian fishermen were shot dead and an equal number were injured. If this situation continued, they would have to leave fishing and take to begging.

In village Vercode, eight fishing boats went to fish as a group near Talaimannar in the first week of May 1998. The Sri Lankan navy captured all the boats, but released four boats instantly. The rest were held in the sea in Talaimannar. For four days they were forced to fish and the catch was taken away. In the same village, fisherman Sahayraj's fishing boat disappeared with its four fishermen when they went fishing in December 1997. Sahayraj had spent around Rs 40,000 to search for them. Sahayraj himself was arrested with his companions in March 1998, when they went fishing near Nedundhevve island. Their belongings were taken away, they were badly beaten up and were released after 24 hours.

Felix Gomez, Assistant Director, Fisheries, in the region of Rameswaram, gave a figure of casualties of Indian fishermen: from 1983 to 1997, 74 fishermen were shot dead and 251 injured in the sea. Fishermen of Rameswaram, Pudukkottai and Nagapattinam districts were suffering heavily. In Nagapattinam district, dozens of fishermen bear marks of injuries in their legs, hands or chests, suffered in various incidents of firing in the sea.

SRI LANKAN FISHERMEN IN INDIAN JAILS

Sri Lankan fishermen are also being caught regularly in the seas by the Indian coast guard and navy. The newspaper *Island* reported on 18 May 1998 that around eighty Sri Lankan fishermen were under detention in India and twenty-five vessels had been confiscated altogether. S. Gautama Dasa, Deputy High Commissioner of Sri Lanka said in Chennai that according to their information, there were more than a hundred Sri Lankan fishermen in Indian jails.

W. Antony Vincent, a Sri Lankan boat owner from Munnakara, Negombo, was desperately waiting outside the Madurai Central Jail in Tamil Nadu in May 1998. He had come to India to get his confiscated boast 'Philip Sahana' and five fishermen released. Vincent had already spent more than Rs 10,000 in reaching Madurai and meeting several people. His five-year-old boat had cost him Rs 9 lakh, out of which he had taken Rs 3 lakh as loan from a bank. He had to repay Rs 15,000 every month. His appearance bespoke desperation. He recounted the event: 'I was not on the boat. So I don't know how the capture had taken place. The captive fishermen—K.S. Nicholas,

W. Wilbert, K.S. Joseph, Sirinimal Fernando and Wijendra Wadugu Chandra—have written to me that they were fishing on our side of the Gulf of Mannar, when the Indian navy came and captured the boat and them. The boat is kept in Keelakarai, south of Mandapan. I want my boat back soon, then only can I survive. I am also trying for the release of the fishermen ... Maybe after some time, I will get rid of all this and shift to some other profession. One day in one country, tomorrow in another country—how can one survive in this way!'

In Madurai Central Jail, we met W. Wilbert. He was returning after catching fish on 3 April 1998, when their boat was followed by the Indian navy or coast guard. When asked to surrender, the fishermen tried to flee. The navy people said that they were being caught because the Sri Lankan navy was killing Indian people. The captives were first taken to Mandapan. They remained there for two days, without being produced before the magistrate. Thereafter, they were kept in the custody of customs for two days in Rameswaram. After four days of arrest, they were produced before a magistrate. The magistrate did not ask them anything, nor did they tell anything. There was nobody to give them any advice what to do. Then they were produced before the magistrate several times. They did not know what their fault was and what cases were imposed against them.

On 25 May 1998 fifteen Sri Lankan fishermen were released from Madurai Jail. They had been charged under the Customs Act for smuggling shark skins! Nothing incriminating had been found in the vessels. A 'confession' was obtained to the effect that shark skins were being brought for sale in India, to obtain a better price. Sri Lanka had a much better price for marine fish than India and shark skin is an internationally traded item for which no significant price differences were likely. In fact, merchant exploitation ensured that India's beach prices were often lower than they should be. Customs at Tuticorin took a decision to levy a penalty of Rs 47,000. They also prosecuted the fishermen. In this miscarriage of justice, genuine fishermen were convicted as common smugglers. According to the Ministry of Fisheries and Aquatic Resources Development, Sri Lanka, in the past five years (1993–7), 269 Sri Lankan boats were arrested for fishing illegally in foreign territorial waters and 338 fishermen were detained.

It took six to twelve months to release a Sri Lankan fisherman, but it took longer to release captured boats. S. Gautama Dass, Deputy High Commissioner of Sri Lanka said that whenever they got any information regarding the arrest of their fishermen, they informed the Tamil Nadu state government. Then the state government got a report from the departments concerned and sent it to the central government. The central government took a political decision about the release. The Sri Lankan mission was not supposed to provide any legal support to the captive fishermen. There also remained a big information gap regarding the whereabouts of the captive fishermen, their arrest and the court sentence. They were thus left in a pathetic and helpless situation.

THE HISTORY OF CONTENTION

The travails of fishermen in the territorial waters between India and Sri Lanka have their genesis in the Maritime Agreements signed on 26 June 1974 and 23 March 1976 between the two countries. The 1974 Agreement demarcated their maritime boundary in the Palk Strait and ceded Kachchativu to Sri Lanka. The 1976 Agreement demarcated the boundary in the Gulf of Mannar and the Bay of Bengal and barred each country's fishermen from fishing in the other's waters.

Kachchativu is a small uninhabited island, situated in the Palk Strait 13 and 16 kilometres from the nearest points of Sri Lanka and India respectively. With cession, article 5 of the 1974 agreement said, 'Subject to the foregoing, Indian fishermen and pilgrims will enjoy access to visit Kachchativu as hitherto, and will not be required by Sri Lanka to obtain travel documents or visas for these purposes.' Article 6 said, 'The vessels of India and Sri Lanka will enjoy in each other's waters such rights as they have traditionally enjoyed therein.'

Not only was this agreement or any of its provisions not discussed with the state governments, fisherfolk organizations or political parties, but it was also not debated in the Indian Parliament before its commencement. There were other dominant considerations between the then heads of states. The aftermath of the 1971 insurrection of Janatha Vimukthi Peramuna (JVP) witnessed a sharp fall in the credibility of Mrs Bandaranaike's government. The country went through a severe economic and political crisis. In the backdrop of this internal scene, the Kachchativu settlement contributed to a large extent in lifting the low morale of the Bandaranaike regime.

The maritime boundary settlement also helped curb the anti-Indian hysteria within Sri Lanka and strengthened the relations between the governments of Indira Gandhi and Sirimavo Bandaranaike. Sri Lanka also extended its support to India on some vital issues. While several countries condemned India's 'peaceful' nuclear explosion in May 1974, Sri Lanka accepted India's stance on using its newly acquired nuclear capability only for peaceful uses. Besides, when Pakistan tried to use the fifteen member ad hoc UN Committee as a forum to attack India over the nuclear explosion, Sri Lanka in its capacity as Chairman of that Committee prevented the forum from doing that. Thus the Kachchativu agreement was a clear political decision, with no thought or consideration to fishermen.

The 1976 Agreement was entitled 'Agreement between India and Sri Lanka on the Maritime Boundary Between the Two Countries in the Gulf of Mannar and the Bay of Bengal and Related Matters'. There was an exchange of letters of the same day between India's Foreign Secretary and Sri Lanka's Secretary to the Ministry of Defence and Foreign Affairs. These letters also constituted an agreement between the two countries. Significantly, this agreement was done when India was under Emergency, and no discussion or dissent was given any space.

Paragraph I of the Exchange of Letters read, 'With the establishment of the exclusive economic zones by the two countries, India and Sri Lanka will exercise sovereign rights over the living and non-living resources of their respective zone. The fishing vessels and fishermen of India shall not engage in fishing in the historic waters, the territorial sea and the exclusive economic zone of Sri Lanka, nor shall the fishing vessels and fishermen of Sri Lanka engage in fishing in the historic waters, the territorial sea and the exclusive economic zone of India, without the express permission of Sri Lanka or India, as the case may be.'

It was the different interpretations of this portion of the Exchange of Letters and of Article 5 in the 1974 Agreement, that led to the controversy whether or not Indian fishermen had the right to fish in and around Kachchativu. The Sri Lankan government argued that the relevant portion of the Exchange of Letters superseded articles 5 and 6 of the 1974 Agreement. It also claimed that article 5 allowed the Indian fishermen only to dry their nets in Kachchativu and not to fish in and around the island.

Since then, representatives of the two countries have met for the umpteenth time to solve the riddle of the Palk Strait deaths of Tamil Nadu fishermen, but have always ended with futile exercises. Professor V. Suryanarayan, Director, Centre for South and South–East Asian Studies, University of Chennai, who has also studied Kachchativu and the problems of Indian fishermen in the Palk Bay Region, put forward the argument that it was always a firefighting exercise by both governments, without removing the causes of the fire. In fact, nobody cared in Delhi and Colombo regarding the shooting and killing of some hundred poor fishermen.

FALLOUT OF ETHNIC CONFLICT AND TERRORISM

Sri Lanka's ethnic conflict, penetration of Tamil militants and LTTE in Tamil Nadu coastal areas, assassination of India's Prime Minister Rajiv Gandhi, and the subsequent catharsis which the government and people of Tamil Nadu underwent have all taken a heavy toll on coastal fisherfolk of both sides in recent years. It is now not only the coast guard or navy, but also the militants, who fire, shoot, arrest, harass or capture the fisherfolk at will.

The immediate fallout of the rise of terrorism and militancy on the coastal fisherfolk was that in 1993 Sri Lanka banned all types of boats in its territorial waters, extending from Trincomalee to Manar, giving a *carte blanche* to its navy to open fire on any unauthorized boats. In times of heightened conflict, Colombo promulgated emergency regulations, which convert its territorial waters into a Prohibited Zone. For reasons of security, Sri Lankan fishermen living in Jaffna and Mannar face restrictions on the type of boats they can own, areas where they can fish, and the time they can spend in the sea. Similarly, the Indian government applied hard measures to prevent the infiltration and movement of militants.

Indian fishermen were a regular target of attack allegedly from the LTTE of Sri Lanka. On 23 April 1997, a number of Indian fishermen were injured when the LTTE made an abortive attack on a major Sri Lankan naval camp at northern Talaimannar, using Indian fishing vessels as cover. The militants virtually dragged the Indian fishing vessels along with them. Murganandan, President, Ramanthapuram District Fishermen Association, said that the LTTE people often take away the fishing boats and send back the crew. The fishermen of Rameswaram had lost at least forty boats in the recent past. As recently as March 1998, LTTE people came near Tondi on the Indian side of the sea and forcibly took away one fishing boat. M. Sahayam of Vercode village voiced similar grievances. Though they were traditional fishermen for generations, they were being thought to be LTTE people or their sympathizers. When LTTE activities or the state's actions against them increased, the killings of fishermen also increased.

There also exist some grey areas regarding the fisherfolk *vis-à-vis* terrorists. It was alleged that some anti-social elements were involved in supplying goods to the militants. There were about thirty mecha-nized boats (owned by people outside the fishing community) which actually did no fishing, but were involved in this illegal activity. The fishermen did not consider these operations as political in nature, but as ways of making quick money.

PALK BAY FISHERIES: A MASSIVE GROWTH

The Assistant Director of Fisheries in Rameswaram region said that if fishermen did not cross the border there would be no fishing in the region. The attraction of fishing in the Sri Lankan waters beyond Kachchativu, upto the Delft island off the Jaffna coast, would be prawns. It is said that the ocean currents and sedimentation on the Sri Lankan side of the Palk Strait have made it a rich field of tiger prawns, which fetch a high price. Another answer lies in the massive growth of fishing activities in the region. The Fisheries Department reports that trawling boats operating from Rameswaram numbered about a thousand. In the traditional sector, there were about 1500 craft. Ram-anthapuram, the district to which Rameswaram belonged, ranked first in the marine fish landings of the state, contributing 23.57 per cent during the 1993–6 period. The growth rate of marine fish land-ings of the district was 44.9 per cent in 1987–90, 10.6 per cent in 1990–3 and 20.7 per cent in 1993–6, compared to 9.9, 8.3 and 9.2 per cent respectively for the whole state. In the coastal region, Palk Bay with only 27 per cent coastline contributed 36.7 per cent of the state's fish landings in 1992–6, while the other two major coasts, Coromandal and Gulf of Mannar, with 35 per cent and 32 per cent respectively of the coastline contributed 28.6 and 25.9 per cent respectively.

The income-sharing system in the trawling boats in Rameswaram also put pressure on the fisherfolk to go for fishing grounds much closer to the Sri Lankan coast, where more shrimp was available. Unlike elsewhere in the state, where net income was shared between the owner and the crew (60:40), here the boat owners paid daily wages to the fisherfolk and extra wages for overnight trips. As an incentive, for every kilogram of shrimps caught the driver gets Rs 20, the second hand Rs 15 and the deck hands Rs 10.

SRI LANKAN DEEP SEA FISHING IN DEEP WATERS

Sri Lanka suffered from overfishing in the coastal waters and to address this, the government encouraged more deep-sea fishing. The country's marine fisheries sector was currently producing around 220,000 metric tonnes per annum, from two-thirds of the area of its economic exclusive zone. It is said that fishery is maximally, exploited if not over-exploited.

A report compiled by Steve Creech with inputs from Sri Lankan organizations like the Forum for Human Dignity, Social and Economic Development Centre, National Fisheries Solidarity and National Union of Fisheries said that there was a rapid increase in the number of mechanized coastal fishing craft, no change in traditional craft, demise of one-day deep-sea vessels and a near-explosion of multi-day fishing boats. According to government figures, in 1995, the country's multi-day fishing fleet was 1543 which is now nearer 1800. The government also announced another round of subsidies to deep-sea boats. However, the deep-sea fishery within the country's EEZ is nearing over-exploitation: 19 per cent of small deep-sea boats were reported to be fishing outside Sri Lanka's territorial waters. Of the larger boats, 39 per cent were fishing beyond Sri Lanka's territorial waters. For boats operating out of Negombo and Chilaw, 74 per cent and 64 per cent were going beyond. Deep-sea fishing boats are also getting bigger, another indication of the desire to fish further from Sri Lanka and for longer.

The report concluded that it was difficult to be optimistic about the future. As long as the Sri Lankan government policy persists in seeking an answer to over-fishing in coastal waters, by promoting deep-sea fishing as an alternative, the problem is likely to continue. The most viable option for many boats is to poach fish from Indian or Maldivian waters, regardless of the risks involved.

The Himalayan Efforts

A Chipko demonstration

Tree planting as a part of Chipko Movement

Trees and Shades
Villages of the Chipko Movement

The Chipko movement has spread far beyond the Chamoli district in Uttar Pradesh where it began in 1973. In several ways, the philosophy of the movement is being carried forward.

Bacher, 20–25 kilometres away from Gopeshwar, the district head-quarters of Chamoli is among the several villages whose lives have been changed by the movement. On the road leading to the village one sees a beautiful nursery built by the villagers and Dashauli Gram Swarajya Mandal (DGSM), a mass organization which is an offshoot of the Chipko movement. Every year Bacher and many other villages around it take their favourite plants from this nursery for planting and are not dependent on the government's social forestry programme.

When the Chipko movement gained momentum in the region in 1973, the demonstrations and dharnas held at various places inspired the people of Bacher, though they were not directly involved in the movement. It was not forest, but the issue of electricity that first drove the villagers' action. The villagers had for long been requesting the authorities to bring electricity to the village. They had also deposited the requisite money. When the electricity board failed to take action, the women of the village took the initiative and called a meeting on the issue. When they contacted the Dashauli Gram Swarajya Mandal (DGSM), the electricity board was given a warning. Within a few months electricity reached the village and the women took out a victory procession.

Next, coming together under their own Mahila Mangal Dal, they focused on the problem of deforestation in the village. The trees in private, governmental, and joint forests of Bacher were indiscriminately

This is a substantially revised version of the paper originally published in *Down to Earth*, 15 June 1992.

and illegally being cut by the government departments. The villagers also, lacking awareness about forest preservation, contributed to the destruction.

The Mahila Mangal Dal made a few rules: they would not allow any cutting of trees in the forests; neither any raw wood nor branches and leaves of trees would be allowed to be broken off. Vigilance would be constantly kept on the forests; every year forest transplantation would be undertaken and people would grow their own gardens. Those who left animals in the open or chopped raw wood would be fined. There would be a guard for the protection of trees who would be paid Rs 500 per month through contributions from every house; members of the Dal would survey the forest twice a month; whichever family needed trees for building a house or for any other purpose would have to take permission from the Dal and pay Rs 50; for equal distribution of fodder during the rainy season one member from every household could cut all the grass; dry and fallen leaves from the trees would be collected and after the rains every family would be given one 'kanda' of fertilizer prepared from it; domesticated animals which destroyed trees were to be kept confined and the women would plant trees of their choice twice a year. By and large, the villagers followed the rules, but the Forest Department, contractors, and some of their local supporters did not follow suit, thereby causing tension.

In one incident, the sarpanch of the village in league with a contractor sold a part of the community forest. Some trees had already been cut when the women learnt about it. The Mahila Mangal Dal annexed the wood loaded in the truck, lodged a complaint against the sarpanch and his brother and gave witness in the court. They also campaigned to defeat him in the elections, and sought to make a woman the sarpanch.

For their outstanding contribution in protecting and planting trees, the Dal received the Indira Gandhi Priyadarshini Vrikshmitra Award in 1984 from the then Prime Minister Rajiv Gandhi. A year later in 1985–6, they were facing governmental threats, arrests, and legal action. It happened when, on the pretext of providing dry and decayed wood to the cities and supplying firewood to Gopeshwar city, the Forest Department ordered the cutting of 1600 trees, including oak and other broad-leafed varieties, from the forest located on the border of Bacher. The Mahila Mangal Dal of Bacher and other villages near it, Tangsa, Siro, Kathud, and Gair, requested the forest conservator of Garhwal Circle to withdraw the order since the region was located on a hill slope and the cutting of trees would lead to soil erosion and landslides. Also, this was the only tract from which the villagers of a dozen villages received grass and dry wood. When their request was rejected, the women declared that they would personally stop the trees from being felled. They were threatened with legal action by the forest and police departments if they went ahead with their decision. When the contractor's labourers came to cut trees the women stopped them. The contractor offered to pay each woman Rs 1000 which they refused and drove away the contractor. The Forest Department even managed to bribe two village men to help in the cutting. The women fought more militantly, and after a long struggle, managed to save those 1600 trees.

The women have also put up fences where required and started farming malta and lemon. One summer night, when the government forest caught fire, the villagers extinguished it by risking their own lives and kept watch the whole night. The people of the Forest Department came to the accident site the next day.

The response of the men to the Dal's activities has changed from initial opposition to gradual acceptance. The sarpanch of the village is a woman as were all the nine members of the *van panchayat*. A woman contested the post of *pradhan*. The village pradhan usually sought the advice of the Dal and the van panchayat in matters relating to rural development schemes and other programmes. The women have turned their attention to other needs such as raising the grade of the village school up to class 12. At present, children wishing to pursue higher education have to go far off or leave studies. Just as the village had gifted land for the nursery, the women have managed to get land for the school sanctioned. The Dal itself is contributing money for the expansion of the school.

TANGSA: WATERSHED DEVELOPMENT

In Tangsa village of Dasholi block there were various tensions regarding interrelationships between different villages, forward and backward caste relations within the village, and the future of industry and employment. In 1983 the Forest Department started afforestation of large parts of governmental land in Tangsa, but the villagers were indifferent and did not participate in the scheme. Then at a meeting of DGSM and the Dal it was decided that the village would run its own afforestation programme.

The Dal started its work in 1991 in earnest. First of all it selected the plants according to the people's needs of fuel, fodder, and fruits. It made certain rules: raw wood would not be cut; animals would not be allowed in the area of afforestation; there would be wall fencing twice a year in different areas; those who cut wood illegally would have to pay a fine of Rs 100, outsiders who encroached on the forest would have their axe confiscated; a guard would be appointed to keep away encroachers and every house would pay Rs 100 per month towards his pay. After two years of hard work, the results of the afforestation drive were visible. The villagers could cut fodder within the village. The women planted trees twice a year, on waste-land, government land and on the roadsides without aid from any outside agency.

Tangsa village lies between Govindi and Balkhala rivers. Between these two rivers lies a large catchment area which had to be protected from land degradation and landslides by planting of trees, develop-ment of ponds, and proper land use. Along with the DGSM, the women chalked out a plan for the protection and development of the catchment area. They selected appropriate land for this purpose and took an oath to plant 300–400 trees every year. They selected plants that would be appropriate for land and water preservation and would also provide a source of economic benefit for the villagers. Walnut trees were planted on two hectares of land. On about seven hectares of degraded land and landslide area they planted malta and lemon. Most of the villagers own land on the landslide area. On these deodar trees were planted.

The malta and lemon crop, after these efforts, is not only sufficient for local consumption but is also sold in the market at Gopeshwar, as are the vegetables grown in the village. Many households keep a milch cow or two which is fed only on locally available fodder. Villagers now have a regular supply of milk and ghee. With the introduction of jersey cows, the number of cows, which used to graze in the open has greatly decreased, from 300 to 140. The villagers have repaid the loan taken for buying the cows. They hope to sell deodar trees as well. Excess fodder available is also sold for about Rs 25,000.

There existed a van panchayat in the village for the last thirty years, but for the first time in the 1990s, a woman was elected its sarpanch. The men's opposition to the women's participation has petered out, but the men still have a weakness for liquor and some young men are on drugs. The caste and class divisions still prevail, and the voluntary programmes of environment conservation and rural development operate, within these limits. The Scheduled Caste families, seeing no concrete economic benefit for themselves in these programmes, had little desire to participate in the joint programmes. They usurped a part of the common land, at times cut and destroyed tree plantation, or cut trees and sold them. Also, the schemes organized under the gram sabha and run according to the priorities of the government were in a way imposed, rather than prioritized by consultation. Programmes like building of maternity homes and panchayat houses were fine, but what would they do about other innumerable government schemes related to roads, irrigation, electricity and water?

SUCCESS AND FAILURE OF THE MOVEMENT

In the Joshimath division of Chamoli district, village Heuna on the Chamoli–Badrinath route was till a few years ago known to be arid and lacking in natural resources. Though every family of this forward-caste village had about 8 to 10 acres of agricultural land with more than 100 acres of private forests, it lacked irrigation resources and faced frequent fodder and firewood crises.

In 1981, some women of the village went to a DGSM camp and were inspired to form a Mahila Mangal Dal. Environmental protection and employment generating developmental works got under way and the whole village became associated with them. Since it was difficult to improve agriculture in the village planting fruit trees was undertaken on agricultural land. Many households planted malta, lemon, banana, and other local fruit trees. Soon enough, the programme was adopted extensively by the whole village with voluntary labour as well as paid employment under a government scheme organized by the Dal. Within a few years enough firewood, fodder, and wood were available within the boundaries of the village.

But prosperity brought its set of problems. The abundant fruit did not have a local market. In 1992, fruits rotted because of glut. There was no infrastructure nearby for fruit preservation or fruit processing. Though the Garhwal Vikas Mandal, a government body for the hilly region, had a scheme for buying a certain amount of malta at remunerative prices, it did not work in Heuna or its nearby areas. The Heuna villagers had to go to the far-off Pipalkoti market where they sold malta-oranges at the dirt cheap rate of Rs 10–12 per hundred. Similarly, the villagers did not know what to do with the large quantity of fodder produced. The women got two bundles of grass daily. Every family had three to six hilly cows or buffaloes which hardly gave half a kilogram of milk each. The struggle to produce fodder was not worth this poor return. High-yielding jersey cows were inaffordable.

The basic aim of Chipko is not merely to protect trees but to provide local opportunities for village employment. In the Garhwal region, however, after the movement started, all forest-related industries closed down. Small-scale projects related to wood, herbs and other forest produce wound up because of a number of restrictions and were not replaced by any new industrial works. One factory, which used to provide regular employment to 25–30 people in forest industries, and had an annual profit of Rs 1 lakh, was shut down and was indebted to the extent of Rs 70,000. This situation alienated even the entrepreneurs who believed in the Chipko principles.

The Chipko movement has six interwoven principles: (a) there should be a ban on the commercial cutting of trees; (b) forests should be surveyed and on the basis of minimum needs of the people, a reorganization of traditional rights should take place; (c) forest should be rejuvenated with the help of local participation and attempts should be made to increase tree cultivation; (d) all kinds of contractorship related to forest activities should end, and instead village committees should be formed; (e) forest-related home-based industries should be developed and for it raw materials, money and technique should be made available; and (f) based on local conditions and requirements, local varieties should be given priority in afforestation. As long as all the principles are not taken into account, the Chipko movement would remain incomplete.

Sisters Cultivate
Tree-Friendly Marriages

Shubha and Rakesh were getting married in the traditional Hindu manner. But this ceremony in the village of Gwaldam in the Himalaya had an extra feature. Shubha gave a sapling to Rakesh. He accepted it gracefully, and with his relatives and villagers, went away and planted it at an appropriate place. He also donated Rs 100 to Maiti Sangathan, an organization of village girls. The money would be deposited in the Maiti bank account to help fund the group's efforts. The marriage ceremony was performed after the tree planting.

Today's tree planting gave an indirect benefit to Shubha's family which is in debt because the arranged marriage to Rakesh has cost her parents a lot in dowry. 'After marriage, I will go to another house and village. The Maiti Sangathan will look after my plant. My parents have spent a lot on my studies and marriage. This plant will produce direct and indirect benefits for them and free me from dues I owe to my parents. The memory will also keep my village green,' Shubha explained.

Various villages of this Himalayan region, with the initiative and efforts of Maiti Sangathan and the enthusiasm of the young girls themselves, have popularized this new custom of planting a tree at the time of their marriage. In this Himalayan region, the word *maiti* carried a special significance. *Mait* means *maika* or 'mother's place', which embraces the whole village in which a girl is born. And by extension, Maiti connotes that which belongs to the mother's place.

Sarita, a resident explained the idea behind the Sangathan, 'All the young and grown-up unmarried girls of the village meet, come together and form a Maiti Sangathan. The group is led by a *badi didi* (elder sister) and the other girls are the *maiti behanen* or maiti sisters.'

Published in *Gemini News Service*, May 1998.

Under the guidance of badi didi, each maiti sister nurses a plant of her choice in her home. When the marriage of any maiti sister is fixed, the other unmarried girls of the Maiti Sangathan look after it. There is a maiti bank account in every village which has a Maiti Sangathan where the money donated by the groom is deposited.

In the Himalayan region, as elsewhere in India, most of the girls live with their parents till the time of their marriage. Parents arrange their wedding and have to often spend huge amounts in dowry. The money spent on her education and upbringing is also thought to be unproductive, yielding no benefits to her family and parents. Maiti inversed this process and the girl leaves behind a ready asset for her parents, which is not only cherished by them as a symbol of their daughter's love, but which also gives the girl a sense of fulfilment for having reciprocated her parents generosity towards her.

In village after village, such as Tangsa, Heuna, and Kandayi, Maiti Sangathan and maiti forests abound. Kalyan Singh Rawat, a teacher at the state inter college at Gwaldum (Chamoli), is said to be the progenitor and inspiration behind this movement. He said that till date, Maiti Sangathan had been formed in more than 1500 villages. In 1997, in May–June, when most marriages take place, 370 maiti trees were planted. In this way, more than five thousand trees have been planted in Chamoli district. Articulating his philosophy, Rawat said, 'In this land of the great Chipko movement, I wanted to amalgamate environmental protection with sacred feelings and traditions. I wanted young girls to feel a consistent and emotional bonding with trees and forests.'

Central Himalaya comprises Garhwal and Kumaon region of Uttar Pradesh state. The region has a fragile ecosystem, with marked differences in topography, vegetation, climate and soil. During the last couple of decades, the natural resources of the Himalaya have been mercilessly exploited, making the hill slopes vulnerable to landslides and soil erosion, thus reducing soil fertility and making the life of hill people exceedingly difficult. The area has also witnessed many catastrophic floods in these years. This has led to varied responses among the people of the region, including those of the Chipko movement.

At Tangsa village, planting of trees was not confined to the occasion of marriage. Every year the women decide on a day for mass tree planting on a piece of barren land in the village. This is done with the consent of the village head. With the income of such a maiti at Tangsa, ten poor maiti sisters were being provided with economic aid in the form of books.

Chandi Prasad Bhatt, one of the prominent leaders of Chipko movement, who lives in Gopeshwar, was all praise for the maiti initiative. He commented, 'Maiti movement shows that traditional beliefs and myths should be reevaluated. Tradition should be reinvented and included in present conservation measures. It is extremely important to understand the role of socio-economic practices of the inhabitants in consonance with their environment.'

Locked Up!

Threat of Mega Projects

For many years the villages of Papriana and Jamak in the Garhwal districts of Chamoli and Uttarkashi had been conscious of looming disaster but their voices were ignored time and again. In August 1991 a massive landslide nearly wiped out Papriana; in October 1991 a devastating earthquake obliterated Jamak. Now when the Tehri dam controversy and the movement against it once again surfaced with even greater force after the earthquake, the experiences of these two villages of the Himalayan region offer important lessons.

That the entire Himalayan zone is geo-dynamically very sensitive. According to the well-known geologist K.S. Vaidiya, 'the much-faulted central sector of the Himalayan arc—Himachal, Garhwal and Kumaon—have remained seismically quiet for quite some time (twenty years) with regard to higher magnitude earthquakes. In other words, this region is a seismic gap which has not been ruptured by big earthquakes for a long time. The stresses are progressively building up inside, as this segment is being strongly pressed and prodded by the fault-lined Aravalli ridge of the northward-drifting Indian subcontinent. The Uttarkashi–Tehri region lying directly in the line of this ridge is, therefore, in a critically stressed condition.'

The knowledge of geotechnical conditions of this region is rather new and there are a larger number of uncertain and unknown factors associated with any kind of construction activity in it. In spite of this, a vast majority of hydroelectric or other big projects (more than 200), that have been built or are being constructed or planned, are located not far from this uncertain zone of the Himalayas. It is, therefore, natural to be apprehensive about the safety of the structures (dams, tunnels, bridges, buildings) and their impact on the people of the

Published in *Economic and Political Weekly*, 4 April 1992.

region, who would be threatened with earthquake, landslide, and land erosion, natural or construction induced.

PAPRIANA

Papriana, in the Dasholi block of Chamoli district, just a kilometre away from the district headquarters at Gopeshwar, is one of the many Chipko villages where issues of forest rights and conservation are very much alive and the Mahila Mangal Dal has been active. Its seventy-plus households are primarily engaged in agricultural works. In spite of being a part of the city area of Gopeshwar, Papriana had no basic facilities like water, electricity, and roads. The pipelines, installed for drinking water supply many years ago, never received any water. Electricity poles and wires were also installed but there was no electricity in the village. Even so, the district administration imposed a water and electricity tax on the villagers. When the villagers refused to pay these taxes, they were declared defaulters by the local administration.

Despite such difficulties, Papriana showed a remarkable degree of initiative in the management of its forest resources. Earlier, fodder and fuelwood used to be extremely meagre in the surrounding areas of the village. The women had to spend six to eight, even ten hours everyday to collect fodder and fuelwood from far-off places. Then came the Chipko movement. A Mahila Mangal Dal was formed in 1979 and it took charge of the afforestation of the surrounding barren land of the village. The Dal also formulated and implemented strong rules for the conservation of forests. The Dal's efforts enabled the village to fulfil its needs of fodder and fuelwood for six months within the village.

The urbanization of Gopeshwar town has also degraded the quality of life in Papriana. The village has become the outlet for sewage water from the town. The sewage water flowed into Papriana village and its agricultural land, causing land erosion and waterlogging. The problem was worsened because the ground water level of Papriana and surrounding areas was always high. During the rains, many areas became waterlogged and many houses were damaged. The villagers demanded a solution to the problem at various levels, including that of the chief minister. The District Magistrate finally took note of the situation in 1981, after the sewage water created havoc in Papriana. He ordered the Public Works Department to construct a proper sewage system, but the PWD asked Rs 1.53 lakh for the works.

The issue thus remained unsolved and Papriana continued to be ruined by the unwanted flow of water. The landslide and rain in August 1991 gave the final touch to the work of destruction initiated by the flow of sewage water.

JAMAK

Jamak village in Uttarkashi district, situated on the hill top adjacent to the Maneri barrage, had a population of nearly 400 people. This barrage on the Bhagirathi river transported the water through a tunnel at Maneri to Uttarkashi for power generation. In the October 1991 earthquake, the entire village was razed to the ground and more than 72 people died and about 200 cattle perished. The disastrous impact of the natural earthquake increased manifold in this seismic zone by the constant weakening and wrecking of the hills, with the ongoing construction works of dams, barrages, tunnels, blasting and other 'developmental' activities.

Jamak and other villages were surrounded by two big hydro power projects, i.e. Maneri Bhali Hydro Power Project, Phase 1 and 2. In Phase 1, a 39-metre high and 127-metre long concrete gravity dam was constructed on the Bhagirathi river near Maneri. Its reservoir had a maximum water level of 1294.50 metres, minimum 1286.50 metres and the tunnel was 8631 metres long. Phase 2 of the project proposes to build a barrage and power house on Bhagirathi river in which the main tunnel is to be 12 to 16 kilometres long.

The people of Maneri, foreseeing their tragic fate, had demanded to shift the village or to stop the disastrous practice of weakening the base, houses and surrounding areas of the village. They had detailed the possible dangers of continuous blasting, digging and other mega construction activities. They organized many protest actions against the tunnel construction. But no precautionary action was taken by the project authorities. According to villagers, a survey had once been conducted by a tehsildar to assess the damage on the village by blasting and other construction activities but nothing came out of that. Due to blasting and tunnel construction, cracks developed in most of the houses in the village. The area and its soil was greatly weakened and land erosion commenced.

+

Had it not been for the big nalas, the huge volumes of waste water, land erosion and waterlogging, Papriana could have been certainly saved. If the severe and continuous blasting, construction of tunnels and big buildings, etc. had been avoided, Jamak and other villages would not have faced the destructive impact of the earthquake to the extent they did.

Glossary

adivasi	tribal
ahimsa	non-violence
alha-udal	legendary heroes of Mahoba, of the twelfth century, about whom ballads are sung
baari	enclosed vegetable garden
bandh	closed
Banwasi Sewa Ashram	a non-governmental organization to support tribals
basti	a settlement or suburb
bazaar	market-place
bhadai	autumn crop
bhajan	devotional songs
bhatta	distiller
bhoodani	cultivable land voluntarily surrendered by landowners in response to the Bhoodan movement led by the Gandhian, Vinoba Bhave
Bhoomi Haqdari Morcha	an organization working to claim land rights
bidi	a twist of tobacco, rolled in a tobacco leaf
bigha	a measure of land, one bigha is equal to about five-eighths of an acre.
bund	embankment
chakka jam	blockade of roads
crore	ten millions
dalit	literally 'oppressed', more specifically members of the Scheduled Castes, or former untouchables of India
dharna	sit-down strike
gair-mozurva	government land
Ganga Mukti Andolan	a movement to liberate the river Ganga

Ghad Kshetra Majdoor	an organization in valley/hill areas in western Uttar Pradesh representing labour
ghat	landing or bathing place at riverside
ghee	clarified butter
gherao	blockading
gobar	cow-dung
goonda	scoundrel
gram panchayat	village council of more than one village, elected by direct voting
gram sabha	sub-set of a panchayat; either coterminous with or sometimes smaller than a revenue village; the smallest unit of electoral democracy
gram sevak	village-level government worker; the lowest unit of the administration; paid a small sum in cash or given a little land, or both
gram swaraj	village autonomy or self rule
hawaldar	policeman
jail bharo	courting arrest and filling up prisons
jan jagriti	raising the consciousness of the people
jatha	a band of people
Jawahar Rozgar Yogna	a poverty alleviation programme of the government
jawan	soldier
jheel	lake, small pond in north India
jungle jamin jan andolan	a movement of forest, land and people
kattha	a measure of land (80 square yards)
khadi	a thick, coarse cotton cloth
Kisan Mazdoor Sangathan	an organization of farmers and labourers
kisan sabha	an organization of peasants and labourers.
lakh	one hundred thousand
lathi	stick
Lok Shakti Sangathan	an organization to empower the rural poor
mahila mangal dal	women's group working for their betterment
mahua	the tree *Bassia latifolia*, and its flower, from which an intoxicant is distilled
maiya	mother
mandir	temple
masjid	mosque
mauja	village site

morcha	to oppose
munshi	clerk
nakedar	forest guard, from the word 'naka' or check post
nala	drain
Narmada Bachao Andolan	The movement protesting against the damming of the Narmada river under the Sardar Sarovar Project
nishkam karmayog	selfless devotion to one's work
numberdar	village headman, by tradition or heredity
pahar	hill
patel	chief of a village, by tradition or heredity; sometimes all rich and powerful persons are also called by this name
patta	title deed
patwari	keeper of village records; paid by the government
pucca	metalled or sealed
puja	devotional service
rabi	the spring harvest
raja or maharaja	king
rasta roko andolan	movement to block roads
roti	bread
samiti	committee
samskar	values believed to be in-born, ritual modes of conduct
sangharsh samiti	organization for struggle
sarkar	government
sarpanch	head of a panchayat, who has quasi-judicial powers in an area coterminous with that of the panchayat
shramdan	voluntary work
tehsil	unit of a district for revenue administration
tehsildar	authority on revenue matters within a tehsil; also, keeper of land records in his area
thanedar	officer in-charge of a police station
tola	a hamlet
Van Shramik Vahini	an organization of forest workers
vidyalaya	school

Vikas Bharti	a non-governmental organization for people's development
vikas pustika	an identity book
Visthapit Mukti Vahini	an organization fighting for the rights of displaced people.
vividh karyakari societies	different working committees
zamindar	landlord
zila parishad	district council

ABBREVIATIONS

BDO	Block Development Officer
CO	Circle Officer
CRPF	Central Reserve Police Force
DM	District Magistrate
IAS	Indian Administrative Service
ICS	Indian Civil Service
IRDP	Integrated Rural Development Programme
NTPC	National Thermal Power Corporation
PAC	Provincial Armed Constabulary
SDM	Sub-Divisional Magistrate
SDO	Sub-Divisional Development Officer

✧

Select Bibliography

In English

A Dossier on All India Fisheries Bandh, Delhi Forum, Delhi, 1994.

Agarwal, Anil and Sunita Narain (eds), *Dying Wisdom: Rise, Fall and Potential of India's Traditional Water Harvesting Systems*, Centre for Science and Environment, Delhi, 1997.

Agarwal, Anil, Sunita Narain, Ajit Chak, Neeraj Labh, V.A. Nambi, Mukul Sharma, and Sangeeta Synghal, *Floods, Flood Plains and Environmental Myths*, Centre for Science and Environment, Delhi, 1991.

An Approach to the National Fisheries' Policy for Fuller Employment and Sustainable Development in the Eighth Plan, National Fishworkers Forum, Delhi, 1990.

Bahuguna, Sunderlal, *Save Himalayas, Soil, Water and Pure Air*, All India Pingalwara Society, Amritsar, 1995.

Basu, N. G., *Forests and Tribals*, Manisha Granthalaya, Calcutta, 1987.

Baviskar, Amita, *In the Belly of the River: Tribal Conflicts over Development in the Narmada Valley*, Oxford University Press, Delhi, 1997.

Behind the Killings in Bihar: A Report on Patna, Gaya and Singhbhum, People's Union for Democratic Rights, Delhi, 1986.

Bhatt, Chandi Prasad, *The Future of Large Projects in the Himalaya*, Pahar, Nainital, 1997.

Bilgrami, K. S. and J. S. Datta Munshi, *Ecology of River Ganges: Impact of Human Activities and Conservation of Aquatic Biota (Patna to Farakka)*, University Department of Botany, Bhagalpur University, 1985.

Bilgrami, K. S., *Bioconservation and Biomonitoring of River Ganga in Bihar (Munger to Kahalgaon)*, Ganga Project Directorate, Bhagalpur, 1993.

Break-Through Despite Break-Up: Kanyakumari March, National Fishworkers Forum, Cochin, 1989.

Buch, M. N., *The Forests of Madhya Pradesh*, Madhya Pradesh Madhyam, Bhopal, 1991.

The Calamity-Prone Central Himalayas, Lok Soochna evam Sahayata Kendra, Chamoli, 1998.

Can the Clock be Turned Back: Delhi Environmental Status Report, World Wide Fund for Nature-India, Delhi, 1995.

Conference of State Forest Ministers, Ministry of Environment and Forests, Delhi, 1996.

Dealing with the Blue Revolution, Programme for Social Action, Thiruvalla, 1994.

Debroy, Vikram, *Food Processing Industries in India*, Ministry of Food Processing Industries, Delhi, 1994.

Sneha, DARP, Society for Integrated Rural Development, Tamilnadu Environment Council, Human Rights Advocacy and Research Foundation, *Dossier on Impact of Aquaculture (Prawn Farms) on Livelihood, Environment and Human Rights*, Chennai, 1994.

Expert Committee on Marine Fishery Resources Management in Kerala, Government of Kerala, Thiruvananthapuram, 1989.

Fernandes, Walter, Geeta Menon and Philip Viegas, *Forests, Environment and Tribal Economy*, Indian Social Institute, Delhi, 1988.

Fishworkers as Prisoners of War: A Fact Finding Team Report on the Arrest and Detention of Indian and Pakistani Fishworkers, Centre for Education and Communication, Delhi, 1998.

Ghosh, D., *A Low-cost Sanitation Technology Alternative for Municipal Wastewater Disposal Derived from the Calcutta Sewage-Fed Aquaculture Experience*, Asian Institute of Technology, Bangkok, 1990.

Government of Kerala: Fisheries Development and Management Policy, Thiruvananthapuram, 1993.

Guha, Ramachandra, *The Unquiet Woods: Ecological Change and Peasant Resistance in the Himalaya*, Oxford University Press, Delhi, 1989.

Hazare, Anna, Ganesh Pangare and Vasudha Lokur, *Adarsh Gaon Yojana: Government Participation in a Peoples Programme*, Hind Swaraj Trust, Pune, 1996.

Hazare, Anna, *Ralegan Siddhi: A Veritable Transformation*, Ralegan Siddhi Pariwar Prakashan, Ahmednagar, 1997.

Kochery, Thomas and Thankappan Achari, *Indian Fisheries: Ecological and Environmental Problems and Resources Depletion*, Programme for Community Organisation, Thiruvananthapuram, 1984.

Kunwar, Shishupal Singh (ed.), *Hugging the Himalayas: The Chipko Experience*, Dasholi Gram Swarajya Mandal, Gopeshwar, 1988.

Kurien, John and T. R. Thankappan Achari, *On Ruining the Commons and the Commoner: The Political Economy of Overfishing*, Centre for Development Studies, Thiruvananthapuram, 1989.

Mumtamayee, C., *Rural Ecology*, Ashish Publishing House, Delhi, 1988.

Nandakumar D. and M. Muralikrishna, *Mapping the Extent of Coastal*

Regulation Zone Violations of the Indian Coast, National Fishworkers Forum, Thiruvananthapuram, 1998.

The Narmada Valley Project: A Critique, Kalpavriksh, Delhi, 1988.

National Fishworkers Forum Dossier on Deep Sea Fishing, Thiruvananthapuram, 1993.

National Fishworkers' Forum: National and Statelevel Unions' Reports, Thiruvananthapuram, 1991.

Pahari, Ramesh, *Dasholi Gram Swarajya Mandal*, Dasholi Gram Swarajya Mandal, Chamoli, 1997.

Pais, H. and C. S. K. Singh, *Face to Face: Active Interventions in Jhabua*, National Labour Institute, Delhi, 1987.

———, *Feedback from Jhabua*, National Labour Institute, Delhi, 1986.

Protect Waters Protect Life, National Fishworkers Forum, Cochin, 1990.

The Public Hearing on Women's Struggle for Survival in the Fisheries, National Fishworkers Forum, Ernakulam, 1995.

Public Interest: An Independent Newsletter on Human Rights and Allied Issues, Calcutta, 1996.

Rai, S. N. and S. K. Chakrabarti, *Demand and Supply of Fuelwood, Timber and Fodder in India*, Forest Survey of India, Dehra Dun, 1996.

Rai, Usha, Mukul Sharma, Sarosh Bana, and Dinesh Kumar, *Call of the Commons: People vs Corruption, Report of a CSE Media Task Force to Ralegaon Siddhi*, Centre for Science and Environment, Delhi, 1991.

Ram, Rahul N., *Muddy Waters: A Critical Assessment of the Benefits of the Sardar Sarovar Project*, Kalpavriksh, Delhi, 1993.

Rao, K. L., *India's Water Wealth*, Vikas, Delhi, 1979.

Report from the Flaming Fields of Bihar, A CPI (ML) Document, Calcutta, 1986.

Saldanha, D., *A Dissenting Report to the Final Report of the Inquiry Committee Appointed for Thane District*, Tata Institute of Social Sciences, Mumbai, 1992.

Samudra: Triannual Report of International Collective in Support of Fishworkers, Chennai, 1996-99.

Sharma, L. T. and Ravi Sharma (eds), *Major Dams — A Second Look*, Gandhi Peace Foundation, Delhi, 1981.

Shiva, Vandana and Gurpreet Kaur, *Towards Sustainable Aquaculture: Chenmmeenkettu*, Research Foundation for Science, Technology and Ecology, Delhi, 1997.

Sinha, Frances, K. A. Srinivasan, Rajiv Kumar Singh, and Viji Srinivasan, *The Blue Revolution: Case-Study of Women in the Inland Fisheries Sector*, Har-Anand Publications, Delhi, 1994.

Solon, Barraclough and Andrea Finger-Stich, *Some Ecological and Social*

Implications of Commercial Shrimp Farming in Asia, UN Research Institute for Social Development, Geneva, 1996.

Surjeet, Harkishan Singh, *Land Reforms in India*, National Book Centre, Delhi, 1992.

Suryanarayan, V., *Kachchativu and the Problems of Indian Fishermen in the Palk Bay Region*, T. R. Publications, Chennai, 1994.

Tindale, Stephen, *Jobs and the Environment*, Socialist Environment and Resources Association, London, 1996.

Urban Sanitation Ecosystem in the Wetlands to the East of Calcutta, Ashoka Foundation, Calcutta, 1992.

Wainwright, Hilary and Dave Elliott, *The Lucas Plan: A New Trade Unionism in the Making?*, Allison and Busby Limited, New York, 1982.

The Wetlands of Calcutta — Sustainable Development or Real Estate Takeover?, World Wide Fund For Nature-India, Calcutta, 1991.

Whither Common Lands?, Samaj Parivartana Samudaya, Save the Western Ghats Movement, Mannu Rakshana Koota, Citizens for Democracy, Karnataka, Dharwad, 1988.

IN HINDI

Bhatt, Chandi Prasad, *Pratikar ke Ankur*, Dasholi Gram Swarajya Sangh, Gopeshwar, 1979.

Bihar: Yaha Kabra hai Loktantra Ki, Samkalin Janmat, Patna, 1991.

Budhani bani Police Chabani, Lok Swatantrya Sanghatan, Hoshangabad, 1988.

Chambal ke Beeharh: Ek Chunauti, Science Centre Gwalior, 1989.

Gupta, Krishna, *Ganga: Ek Prakitik tatha Sanskritic Dharohar*, Himalaya Seva Sangh, Delhi, 1991.

Hemant and Ranjeev, *Jab Nadi Bandhi*, JaiPrabha Adhyayan evam Anusandhan Kendra, Madhupur, 1991.

Larat Ja Re, Kisan Adivasi Sanghatan, Hoshangabad, 1991.

Larenge — Jitenge: Hoshangabad Zile ke Visthapito ki Larai, Kisan Adivasi Sanghatan, Hoshangabad, 1997.

Lokshakti: Samajik Parivartan ke liye Pratibadha Lok Shakti Sanghatan ka Mukhpatra, Madhubani, 1998-99.

Mishra, Anupum, *Aaj bhi Khare hai Talab*, Gandhi Shanti Pratisthan, Delhi, 1993.

Mishra, Dinesh Kumar, *Badh se Trasth — Sinchayi se Pasth*, Samta Prakashan, Patna, 1990.

Narendra and Birendra, *Eka Aiso Banai*, Kisan Adivasi Sanghatan, Hoshangabad, 1988.

234 • *Select Bibliography*

Pahar 1, 2, 3, 4, Peoples Association for Himalaya Area Research, Nainital, 1983–1989.

Sahi-Sahi Sahitya Sankalan, Sahi-Sahi Prakashan, Bhagalpur, 1981.

Sharma, Mukul, *Bihar ka Diara Kshetra*, Bhagat Singh Research Samiti, Ludhiana, 1988.

Vikas ki Bali Charte Vanvasi: Suvarna Rekha Pariyojana se Visthapan aur Visthapito ke Sangharsh ki Kahani, Visthapit Mukti Vahani, Singhbhum, 1991.

Yogendra and Safdar Imam Kadari (eds), *Ganga ko Aviral Bahana Do*, Ganga Mukti Aandolan, Bhagalpur, 1990.